PROTOPLASMATOLOGIA

HANDBUCH DER PROTOPLASMAFORSCHUNG

BEGRÜNDET VON

L. V. HEILBRUNN · F. WEBER
PHILADELPHIA GRAZ

HERAUSGEGEBEN VON

M. ALFERT · H. BAUER · C. V. HARDING · W. SANDRITTER · P. SITTE
BERKELEY TÜBINGEN ROCHESTER FREIBURG I. BR. FREIBURG I. BR.

BAND VI

KERN- UND ZELLTEILUNG

A

THE CHROMOSOME COMPLEMENT

1968
SPRINGER-VERLAG
WIEN · NEW YORK

THE CHROMOSOME COMPLEMENT

BY

B. JOHN and K. R. LEWIS
BIRMINGHAM OXFORD

WITH 87 FIGURES

1968
SPRINGER-VERLAG
WIEN · NEW YORK

ISBN-13: 978-3-211-80881-8 e-ISBN-13: 978-3-7091-5781-7
DOI: 10.1007/978-3-7091-5781-7

LIBRARY OF CONGRESS CATALOG CARD NUMBER: 55-880

TITEL-NR. 8736

Protoplasmatologia
VI. Kern- und Zellteilung
A. The Chromosome Complement

The Chromosome Complement

By

Dr. BERNARD JOHN
Department of Genetics, The University, Birmingham, England

and

Dr. KENNETH R. LEWIS
Botany School, The University, Oxford, England

With 87 Figures

Contents

Introduction

Biological systems express and reveal their specificity in different ways and at different levels. This specificity may be manifest in the choosing of a mate or a molecule and in the compatibility of a graft or a pollen grain. It is seen also in nutritional requirement and excretion product, in disease resistance and host range, in ecological preference and geographical distribution. It is expressed in chromosome pairing and the clotting of blood, in gametic fusion and in mating-call. All these properties express the distinctiveness of living systems and must ultimately rest on the shapes and surfaces of the macromolecules which contribute to their organisation. These chemical contours, in turn, are determined by the kind, the order and the degree of polymerisation of the monomer components.

Development reveals and allows heredity while heredity prepares and provides for development. Both are expressions of the same specificity which rests ultimately in nucleic acid because, so far as is known, no other material can serve both autocatalytic and heterocatalytic functions. Thus not only is it unique in its capacity for self-replication but its base sequence determines the specificity of proteins. And enzyme proteins are immediately responsible for the peripheral metabolism which enables the organism to impose its own kind of order on the raw materials it absorbs.

The course of development is determined not only by the nature of the genetic material but by its over-all amount and the relative frequency of the different functional units. Differential rates of epigenetic activity matter also. In theory, therefore, differential development within or even between individuals could be determined by the differential replication of the various genetic elements or by their differential activity. And further variation could arise by the differential transmission of these elements between cells. Indeed it would appear that all these possibilities are exploited by living systems.

If like is to beget like, however, any genetic change which occurs during development must be undone, or else germinal units preserved from change must be set aside. As far as is known, genetic changes, even those involving only quantity or relative amounts, are reversible to only a very limited extent so that a change once done cannot be undone. Consequently genetic changes during the development of presumptive germ-lines are either non-existant or minor and confined to a small class of unaggregated determinants. Thus in *Rana pipiens* following nuclear transplantation into experimentally enucleate eggs the male germ cell nuclei promoted cleavage in 43% of the recipient eggs. And 40% of the completely cleaved eggs developed into normal tadpoles (Smith 1965). By contrast, nuclei of somatic endoderm cells promoted cleavage in only 18% of the recipient eggs and none of these developed normally.

Now the number of different determinants required for the genesis of even the simplest living system is large and the behaviour of this host must be co-ordinated in both development and reproduction. The needs of heredity are fourfold. First, the various elements must replicate accurately. Secondly, they must do so more-or-less simultaneously and at the same rate. Third, each determinant must segregate from the product of its replication and fourth, the segregating units must form two groups each containing one representative of each segregating pair. The difficulty of meeting these qualitative and quantitative, temporal and spatial requirements when a large number of determinants are involved is intensified by the very small size of the individual genes. The problems imposed by number and size are both reduced by aggregation and, even in the simplest organisms, the vast majority of the genetic elements are associated into a single, essentially-linear structure. This linear aggregation or linkage of genetic elements substitutes a certain and rigid mechanical form of distribution in place of the uncertainty inherent in a random segregation of large numbers of dissociated elements. Some genes are not incorporated into such linear aggregates and, significantly, both their rates of reproduction and the incidence of their transmission is often erratic. Yet others exist in alternative states in which they may or may not be included in the major aggregate. In the former condition their behaviour is regular but when detached it is not.

In the viruses and the bacteria the gene aggregate appears to consist of little more than nucleic acid. But in plants and animals it forms part of a more massive and heterogeneous organelle called a chromosome. And in the vast majority of these organisms most of the genes are distributed between two or more—usually many more—chromosomes. It is true that the chromosome is not the exclusive mechanism of genetic transmission even in these higher organisms. Here too certain genes are extra-chromosomal and the differential replication and transmission referred to earlier is, for the most part, confined to certain of these.

Chromosomes are longitudinally differentiated structures of varying lengths and some, at least, of this differentiation is microscopically visible. And it is with this visible variation in chromosome number, size and structure that we shall be concerned in this monograph. The principal property of the chromosome complement is clearly constancy of genotype but it is subject to both phenotypic and genetic variation in both development and heredity. Let us begin, however, by considering those features which enable one to distinguish chromosomes and the complements they compose.

I. The Elements of Chromosome Organisation

In discussing the chromosome complement we are concerned with chromosome phenotype. We may therefore characterise a complement in terms of the morphology and the behaviour of its members. These two properties, in turn, may be defined in terms of the nature of, and the variation to be found in, the five discrete organelles which go to make up a chromosome. As a necessary preliminary to an understanding of the chromosome complement we need to consider these components in some detail.

1. The Chromonema

This is the basic chromosome thread which maintains the essential organisation of the chromosome throughout all phases of the cell cycle. One of the most striking aspects of mitosis and meiosis is the physical change which this chromonema undergoes. Measurements of the changes in the length of the chromonemata during mitosis in "living" endosperm of *Haemanthus katherinae* and *Leucojum aestivum* show that they begin to decrease rapidly in length before the dissolution of the nuclear membrane and then continue to decrease until metaphase (BAJÉR 1959). During metaphase there may be a short period when they do not change in length but they start to decrease further before the start of anaphase and continue the process during their separation. Notice that homologous chromosomes may have slightly different lengths in different cells at corresponding stages and even in the same cell. The volume of the chromosome, however, increases until metaphase when it reaches a maximum and then begins to decrease before the start of anaphase, continuing the process through it.

Both the length and the volume changes depend, in part at least, upon a process of internal coiling though, as we shall see (pg. 7), not all parts of the chromonema are equally or synchronously affected. Coupled with this spiralisation cycle the chromosomes undergo a parallel cycle of chromaticity which increases to a metaphase maximum and is then reversed again at telophase. Whether the cycle of spiralisation also accounts adequately for the cycle of chromaticity is not clear.

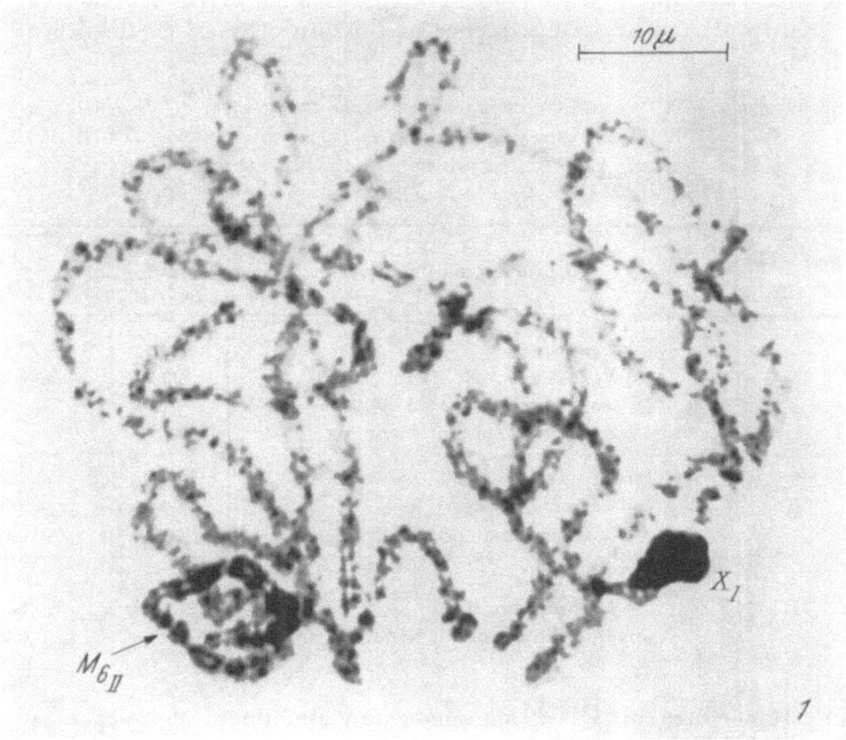

Fig. 1. Chromomeric organisation at zygotene in the grasshopper *Chorthippus parallelus*. Note the pronounced difference in chromomere size between the megameric M_6 bivalent (arrow) and the other autosomes.

At first meiotic prophase the chromonema develops a chromomeric organisation. As the meiotic chromosomes shorten and thicken the chromomeres diminish in number and increase in size as they merge into one another (Fig. 1). The pattern of chromomeres is characteristic for a chromosome and as constant as it. Thus in a number of plant species the chromomeres exhibit a definite gradient at pachytene. They are large on both sides of the centromere and decrease in size as they approach the chromosome ends (Lima-de-Faria 1956). Again in *Zea mays* the modified chromomere pattern produced by a 5–6 interchange has persisted unaltered during the eighteen years which have intervened since this structural mutation was first produced by McClintock (Lima-de-Faria and Sarvella 1962). The chromomere pattern is especially clear in lampbrush chromosomes (Callan and LLOYD 1960) and it is now generally agreed that each chromomere represents a coiled portion of the chromonemal thread. It is also agreed that the chromatic bands of polytene chromosomes are equivalent to the sets of chromomeres found in meiotic chromosomes. Now, the rate of metabolic activity is high both in oocytes and in tissues with polytene chromosomes. In the latter this metabolism is markedly periodic, different loci being active at different stages of development. This activity is associated with the extension of the bands, to form so-called puffs, bulbs and rings, and their subsequent regression. The appearance of the chromosome may be greatly altered as a consequence of this. In oocytes, on the other hand,

Table 1. *A Comparison of Chromomere Characteristics at Pachytene of Meiosis and at Second Microspore Mitosis in Ornithogalum virens.*
Each value represents an average from three cells.
(Data of LIMA-DE-FARIA, SARVELLA, and MORRIS 1959.)

Chromosome No.	Chromosome Character	Stage	
		Pachytene	P. G. M. II
I	(a) Length (μ)	109.1	22.4
	(b) Chromomere No.	105.3	24.3
	(c) Average distance (μ) between consecutive chromomeres	1.03	0.92
II	(a) Length (μ)	89.0	22.9
	(b) Chromomere No.	80.6	23.3
	(c) Inter-chromomeric distance (μ)	1.10	0.98
III	(a) Length (μ)	98.5	20.6
	(b) Chromomere No.	91.0	21.3
	(c) Inter-chromomeric distance (μ)	1.08	0.97

the chromomeres of all chromosomes are functional more-or-less simultaneously. Here the approach to maximum synthetic activity is associated with the development and extension of lateral loops from the chromomeres which undergo a corresponding reduction in size. Likewise, cessation of activity is accompanied by loop regression. Thus, while chromomeric organisation is a chromosome characteristic it is also subject to variation.

Again, chromomeres may be developed at different levels of spiralisation and they are not strictly comparable at different stages. Thus in *Ornithogalum virens* the three chromosomes which constitute a haploid complement can be identified at all stages. The same type of chromomere pattern is found at pachytene and at the prophase of the second pollen grain mitosis (LIMA-DE-FARIA, SARVELLA, and MORRIS 1959). The three chromosomes are four to five times shorter at this mitosis than at pachytene but their chromomere number is also approximately four times lower since chromomere number diminishes in proportion to length in all the chromosomes (Table 1).

The chromonema most usually appears as a single stranded structure, a structure referred to as a chromatid. There is evidence, however, that it is in reality a compound unit. This evidence takes three principal forms:

(1) There have been repeated claims to have seen half-chromatids in mitotic chromosomes. Despite criticisms that these sub-units are artifacts the reality of their existence is borne out by observations on the structure of c-mitotic chromosomes in *Endymion non-scriptus* following treatment with 8-oxyquinolene (Giménez-Martin, López-Sáez, and González-Fernández 1963) and especially by Bajér's (1965) study on unfixed and unstained endosperm chromosomes of *Haemanthus.*

(2) By subjecting metaphase chromosomes isolated from neutral formalin fixed roots of *Vicia faba* to trypsin treatment Trosko and Wolff (1965) have given convincing evidence that each chromatid consists of two half-chromatids. There is even a suggestion that quarter-chromatid units may occur in this species.

(3) Irradiation of late prophase or metaphase chromosomes leads to the development of side arm bridges both at mitosis and meiosis. A number of authors have concluded that at least some of these bridges involve half-chromatid units (see discussion in Peacock 1964).

There seems little doubt, therefore, that the chromonema is a compound unit. A chromosome is usually composed of either one or else two such compound units though in special cases the chromosome may be polytene consisting of many chromatids (see pg. 28).

2. Heterochromatic Regions

The cycle of condensation which the chromonema undergoes during nuclear division is not necessarily synchronous in all members of a complement nor indeed in all regions of individual chromosomes. A standard cycle of condensation—a eucycle—is recognised with an interphase minimum and a meta-anaphase maximum. Certain chromosomes or chromosome segments may, however, depart from this eucycle in various ways and they are consequently described as allocyclic. Allocyclic segments may be more or less condensed than eucyclic ones at particular phases of the cell cycle (Fig. 2). In consequence they stain differentially and thus exhibit heterochromasy as well as heteropycnosity. Allocyclic segments may be either positively or negatively heteropycnotic. In the former case they appear overcondensed and heavily staining while in the latter they seem undercondensed and lightly staining (Fig. 3).

Chromosomes or chromosome regions which show allocycly are described as heterochromatic and are distinguished from the euchromatic and eucyclic regions. The original term heterochromatin was, in fact, borrowed from the description of the sex-chromosomes as heterochromosomes because they stained differentially at meiosis in contrast to the autosomes. The evidence, as it stands, supports the conclusion that while heterochromatic regions differ from euchromatic ones in their behaviour they show only minor differences in fundamental structure (Brown 1966). Thus euchromatic-heterochromatic conversions are well known in cases of position effect (HANNAH 1951, BAKER 1953, and see pg. 144).

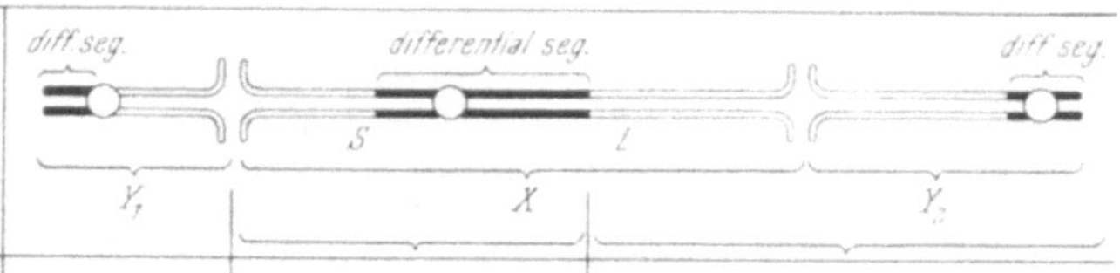

Stage	Y_1	X	Y_2
1. Spermatogonia			
a Prophase	±	+	±
b Metaphase	±/+	–	±
2. Spermatocytes			
a Early diplotene	+	+	±
b Late diplotene to Anaphase-I	–	–	±
c Anaphase-II	–	–	±

Fig. 2. Patterns of heteropycnosity in the sex chromosomes of the marsupial *Potorous tridactylus* (based on SHARMAN and BARBER 1952).

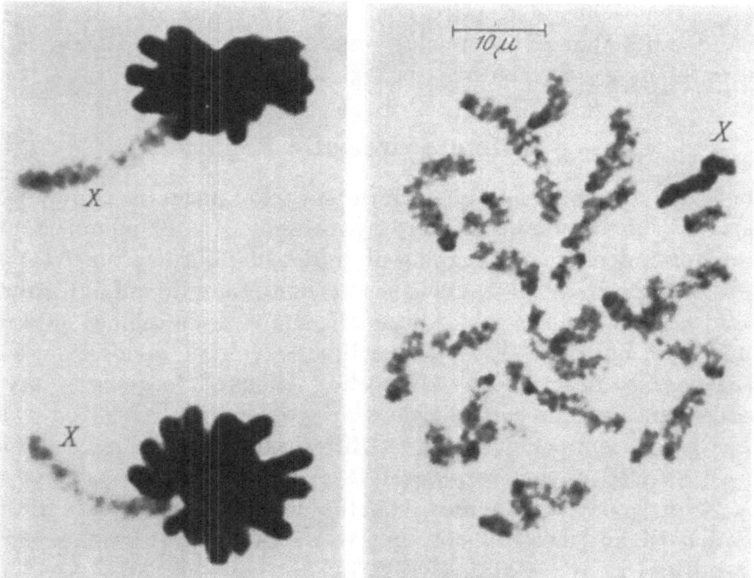

Fig. 3. Negative (telophase, Fig. 3 *a*) and positive (prophase, Fig. 3 *b*) heteropycnosity in the pre-meiotic mitoses of the grasshopper *Eyprepocnemis plorans*.

It will be appreciated that, owing to allocycly, the size relations between chromosomes and the arm-ratios within them may vary considerably with the stages of the division cycle. And both the nature and the extent of allocycly itself may vary owing to both internal and external factors.

For example, allocycly in the form of negative heterochromasy can be created by cold treatment in a variety of plants and animals (Table 2, Fig. 4). These differential or H-segments appear to arise by a process of localised uncoiling (WOODWARD and SWIFT 1964). The degree of differential contraction varies over a wide range. Thus in *Paris japonica* (DARLINGTON and LA COUR 1940) and *Trillium undulatum* (DARLINGTON and SHAW 1959) H-segments may be apparent even at normal temperatures while those of *Trillium stylosum* are not easily distinguishable even after chilling. Further, it would appear that different segments may require different amounts of

Table 2. *The Occurrence of Cold-sensitive Heterochromatic Regions.* (After DYER 1963.)

Genus			Reference
Plants	Monocots	1. *Fritillaria*	DARLINGTON and LA COUR 1940, 1941
		2. *Hordeum*	MECHELKE 1955
		3. *Paris*	DARLINGTON and LA COUR 1938
		4. *Secale*	WILSON and BOOTHROYD 1941
		5. *Trillium*	DARLINGTON and LA COUR 1940, 1941
		6. *Tulbaghia*	DYER 1963
	Dicots	1. *Adoxa*	GEITLER 1940
		2. *Cestrum*	DYER 1963
		3. *Vicia*	LA COUR 1951
Animals		1. *Ambystoma*	CALLAN 1966
		2. *Bufo*	WICKBOM 1945
		3. *Mecostethus*	CALLAN 1942
		4. *Rana*	WICKBOM 1945
		5. *Triturus*	CALLAN 1942

treatment for their visualisation (BOOTHROYD 1953). DYER (1964) has evidence to suggest that the degree of differential contraction may be correlated with the overall amount of H-material in the chromosome arm and in the whole complement (Fig. 5).

Now while cold-treatment causes under-contraction of the H-segments it may have the opposite effect on the euchromatic regions. But this is not uniform. Thus, at low temperature (2° C) purely euchromatic arms in *Trillium grandiflorum* (A and B long arms, A and C short arms) are over-contracted to about 90% of the length at 20° C (Table 3). The E-segments of arms with H-regions (C long arm and D short arm), however, do not show this effect to the same extent if at all.

SMITH (1965 a) has described a comparable differential effect following colchicine treatment in *Chilocorus.* The complement of most species in this genus includes autosomes of two kinds. First, those in which both arms are euchromatic (E : E) and second those which consist of one euchromatic and one heterochromatic arm (E : H). In untreated spermatogonial metaphases all the chromosomes appear metacentric. Following short periods of colchicine treatment, however, the E : H chromosomes become markedly asymmetrical, one arm appearing 2–3 times longer than the other. It is claimed that this effect results from the super-contraction of the E arm but a comparable effect does not materialise in E : E chromosomes. The asymmetry of E : H chromosomes is less pronounced after prolonged colchicine

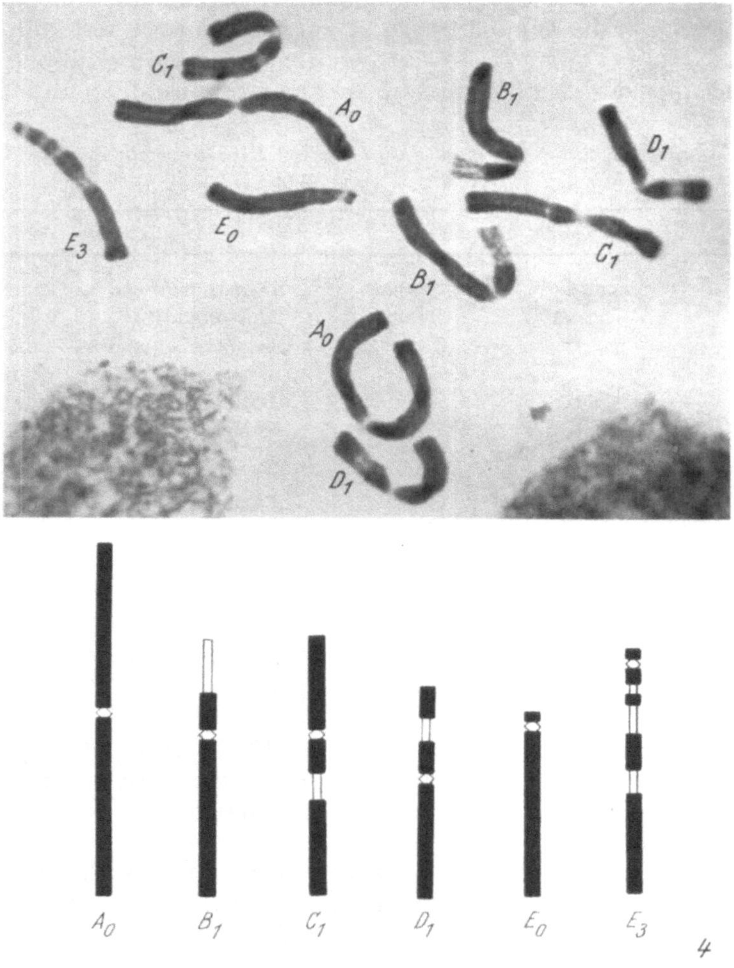

Fig. 4. Metaphase of a chilled root tip in *Trillium grandiflorum* with a chromosome complement of the type $A_0A_0B_1B_1C_1C_1D_1D_1E_0E_3$ (photograph kindly supplied by Prof. ALFRED RUTISHAUSER).

treatment. The neo-Y (E : H) chromosome of *C. hexacyclus* is affected in the same way as E : H autosomes while the E : E X-chromosome behaves like the E : E autosomes. On SMITH's interpretation intra- and inter-chromosome condensation differentials are required. It is not clear, however, how SMITH is able to tell whether it is the super-contraction of E- or H-arms which causes the asymmetry. If, in fact, the latter are involved then, clearly, only inter-segment differences are required.

Table 3. *The Lengths of Heterochromatic and Euchromatic Segments at 2° C Compared with the Total Chromosome Length at 20° C in an Individual of Trillium grandiflorum Heterozygous for H-pattern in Chromosomes C and E.*

Each value is the mean of 10 determinations. E and H are euchromatic and heterochromatic segments while S and L distinguish the short and long arms. (Data of DYER 1964.)

Length (μ)	Chromosome type and arm														Total
	A		B		C 1		C 2		D		E 1		E 2		
	S	L	S	L	S	L	S	L	S	L	S	L	S	L	
E – 2° C	11.1	12.1	3.6	11.1	8.2	9.8	7.6	10.0	6.0	8.2	1.1	12.0	1.1	12.6	166.6
H – 2° C	—	—	3.5	—	—	0.9	1.5	0.9	1.8	—	—	0.6	—	2.1	16.6
Total 2° C	11.1	12.1	7.1	11.1	8.2	10.7	9.1	10.9	7.8	8.2	1.1	12.6	1.1	14.7	183.2
Total 20° C	13.8	14.8	5.6	12.9	8.9	10.7	8.9	10.7	6.3	9.3	1.3	15.1	1.3	15.1	197.4

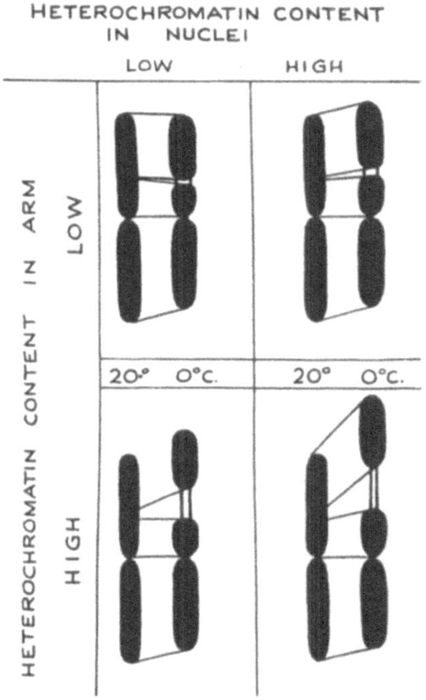

Fig. 5.

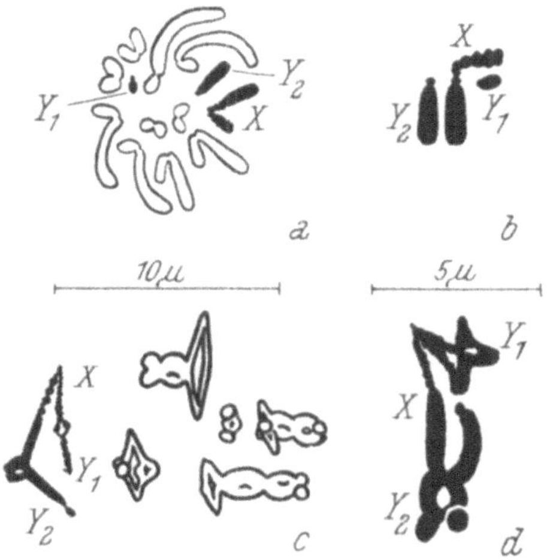

Fig. 6.

Fig. 5. The effect of the amount of heterochromatin in the nucleus and in the chromosome arm of species of *Trillium* on the response of eu- and hetero-chromatin to 4 days cold treatment at 2° C (after DYER 1964).

Fig. 6. A comparison of the structure of the sex chromosomes of *Potorous tridactylus* at mitosis (*a* and *b*) and meiosis (*c* and *d*). Notice that while the Y_1 is less than half the length of the short arm of the X at mitotic metaphase (Fig. 6 *b*) these same segments appear to be of nearly equal length at diplotene-metaphase I (after SHARMAN and BARBER 1952).

Similar variations in arm ratios, and hence relative chromosome size, can arise as a result of differential allocycly between cell types. Thus in *Potorous tridactylus* nearly all the Y_1 is paired at diplotene with the short arm of the X-chromosome so these two regions appear of equal length. But at spermatogonial metaphase the short arm of the X, together with the proximal region of the long arm, is negatively heteropycnotic and it appears to be twice the length of the isopycnotic Y_1 (Fig. 6).

Within a division cycle also, length relations may vary within and between chromosomes as isopycnotic regions approach the degree of condensation achieved earlier by those which condense precociously. This effect has been studied in detail at meiosis by SMITH (1952) in *Tribolium* and BROWN (1949) in tomato (see JOHN and LEWIS 1965). Furthermore, even homologous chromosomes and sets of homologues within a nucleus may be subject to differential condensation (see pg. 58). Thus heteromorphism for the number, the size and the position of cold-induced H-segments has been found in some (*Cestrum*), many (*Fritillaria*) or all (*Tulbgahia*) of the chromosome pairs containing H-segments (DYER 1963) though most of these appear to depend on heterozygosity.

3. Kinetic Organelles

There is in each chromosome a system which regulates chromosome orientation and movement in relation to the spindle. This system may take a variety of forms the commonest of which is the localised kinetic unit or kinetochore. In such systems the kinetic regions do not coil during the mitotic cycle and consequently they appear as non-staining regions, the so-called primary constrictions. These provide one of the most useful markers for describing the chromosome complement because, while their position is constant for a chromosome, it may vary between them. The kinetochore may be localised anywhere along the length of the chromosome but for descriptive purposes three principal chromosome types may be distinguished:

(a) Telocentric—where the kinetochore is strictly terminal so that the chromosome is one armed [1].

(b) Metacentric—where the chromosome is composed of two arms whose relative lengths approach unity.

(c) Acrocentric—where the kinetochore is interstitial but the two arms are very unequal.

On this basis the distinction between meta- and acro-centric chromosomes is clearly a relative one. Because of this LEVAN, FREDGA, and SANDBERG (1964) have proposed a more elaborate system of nomenclature where the position of the centromere is defined in terms of the difference (d) or the ratio (r) between the respective lengths of the long (l) and short (s) arms (Fig. 7). Unfortunately the application of this system is complicated by the fact that the distinction between acrocentric and telocentric chromosomes is not easily made (JOHN and HEWITT 1966 and see pg. 77).

---

[1] This is a new definition of an older term (see pg. 77).

In some species with localised kinetochores all the chromosomes are of the same centric type while in others the karyotype is composed of mixtures of different types (Fig. 8). Complements vary considerably with regard to the extent and distinctness of the primary constriction. In some cases it is very long and, especially in squash preparations, may appear as a discon-

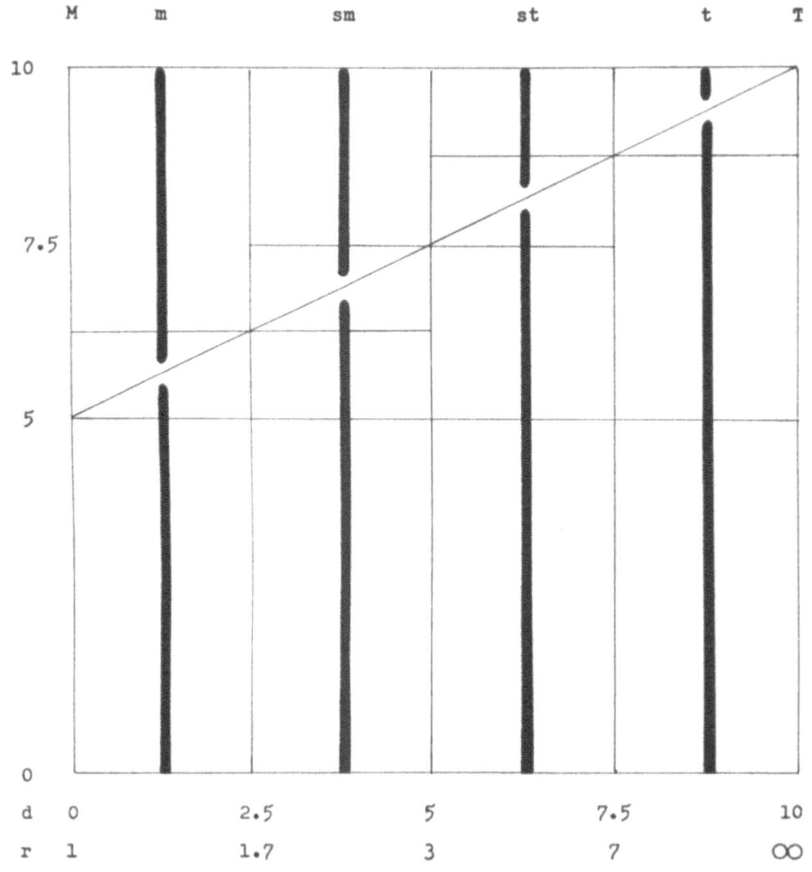

Fig. 7. Variation in the position of the kinetochore considered in relation to four segments of equal length in a chromosome 10 units long (after Levan, Fredga, and Sandberg 1964).

tinuity. In others the position of the kinetochore is marked by little more than a slight decrease in chromatid diameter. A comparable difference is sometimes found between cell types and it may be affected by the over-all degree of condensation. Mitotic inhibitors frequently affect the degree of contraction and also the distinctness of constrictions. In *Pellia epiphylla*, however, even c-mitotic inductors have little affect on the visualisation of the centromeres whose positions can be detected clearly only then they are mechanically active (Fig. 9).

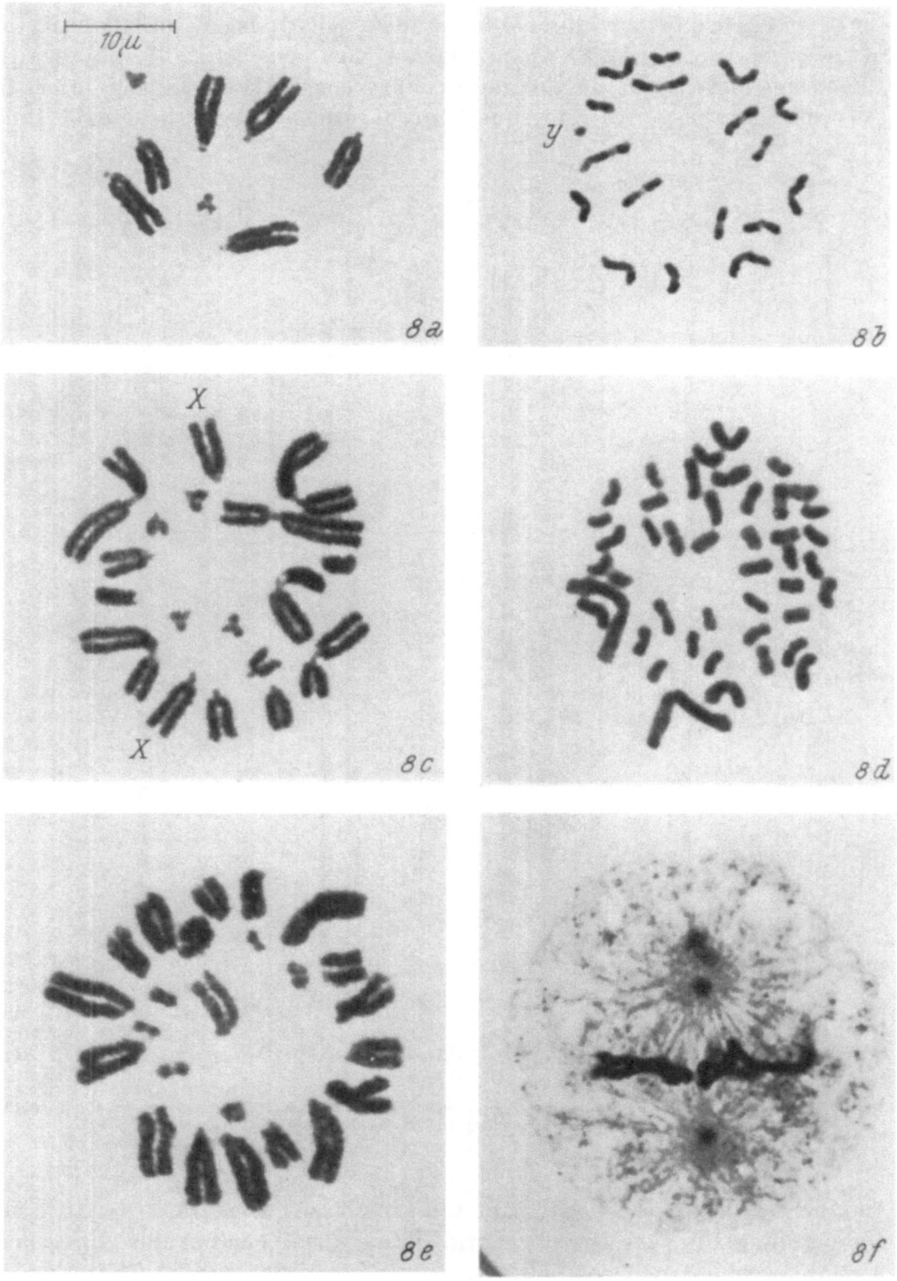

Fig. 8. Mitotic karyotype variation in animals (Figs. 8*a*—*f*) and plants (Fig. 8*g*—*l*).

Animals (all × 3000)

a — *Dugesia lugubris* type E (Platyhelminthes), 2n = 8

b — *Dermestes maculatus* (Coleoptera), 2n = 16 (14 + Xy)

c — *Myrmeleotettix maculatus* (Orthoptera), ♀ 2n = 18 (16 + XX)

d — *Cepea nemoralis* (Mollusca), 2n = 44

e — *Schistocerca cancellata* (Orthoptera), ♀ 2n = 24 (22 + XX)

f — *Parascaris equorum* (var. bivalens) (Nematoda), 2n = 4 compound polycentric chromosomes

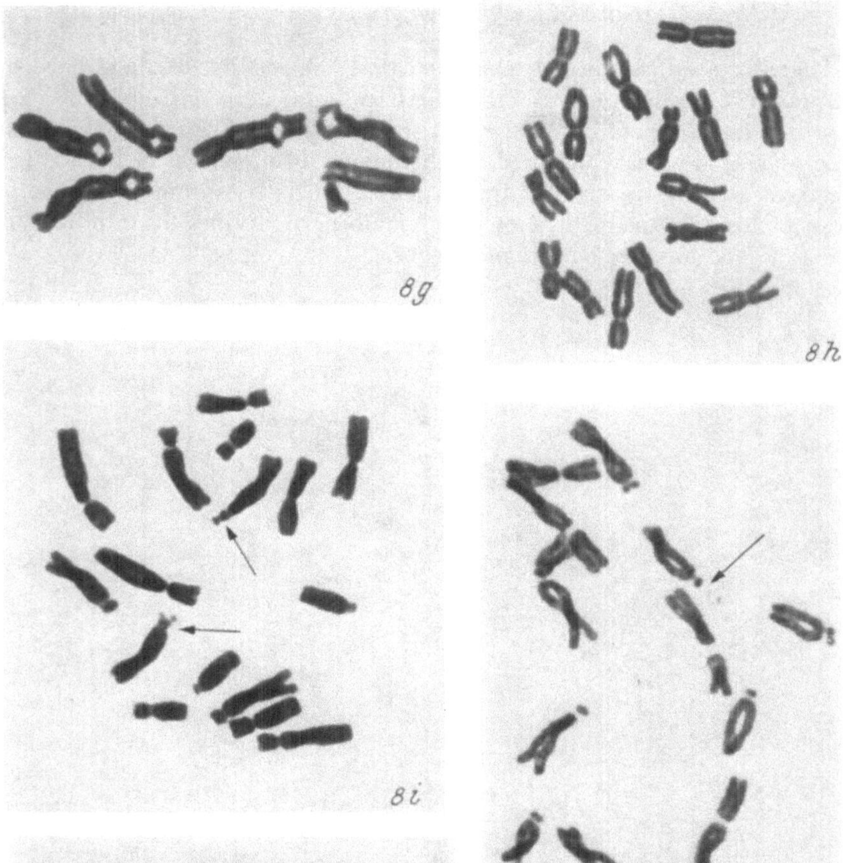

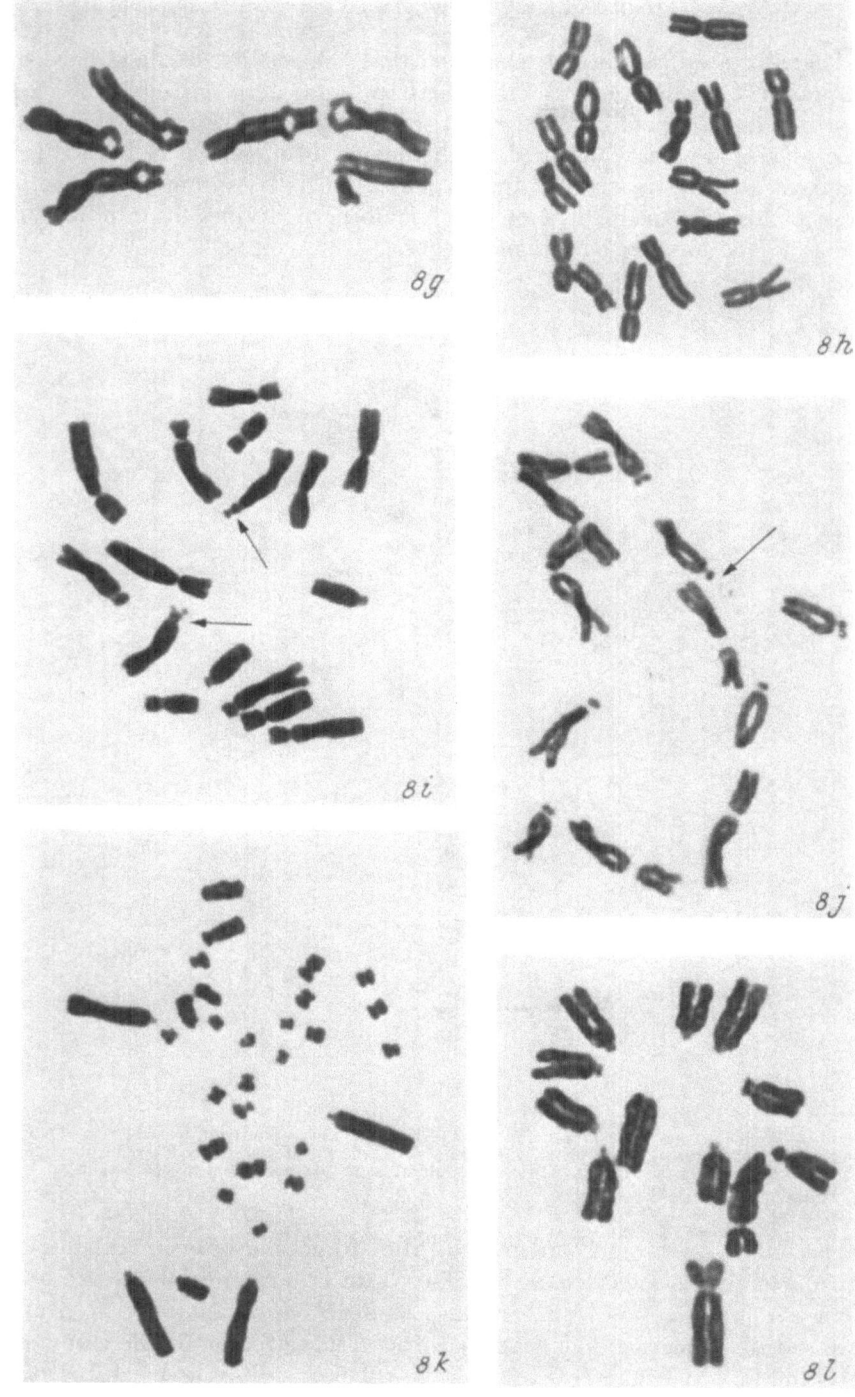

Plants (all Angiosperms)

g — ***Crocus balansae***, 2n = 6

h — ***Allium schoenoprasum***, 2n = 16

i — ***Scilla scilloides***, 2n = 16. Note satellited chromosomes (arrows; photograph kindly supplied by Dr. G. H. JONES)

j — ***Callisia fragrans***, 2n = 12. Note heteromorphism (arrow) for satellited chromosome (photograph kindly supplied by Dr. KEITH JONES)

k — ***Eucomis zambesiaca***, 2n = 30 (photograph kindly supplied by Dr. KEITH JONES)

l — ***Callisia elegans***, 2n = 12 (photograph kindly supplied by Dr. KEITH JONES)

There is now convincing evidence that the localised kinetochore has a compound organisation. In metacentrics and acrocentrics this can be described as a tandem reverse repeat (LIMA-DE-FARIA 1956) since it consists of three principal zones composed of fibrillae or chromomeres or both (Fig. 10). The most striking feature of this organisation is its symmetry for a plane passing through the centre of the kinetochore divides it into two parts related to one another as mirror images.

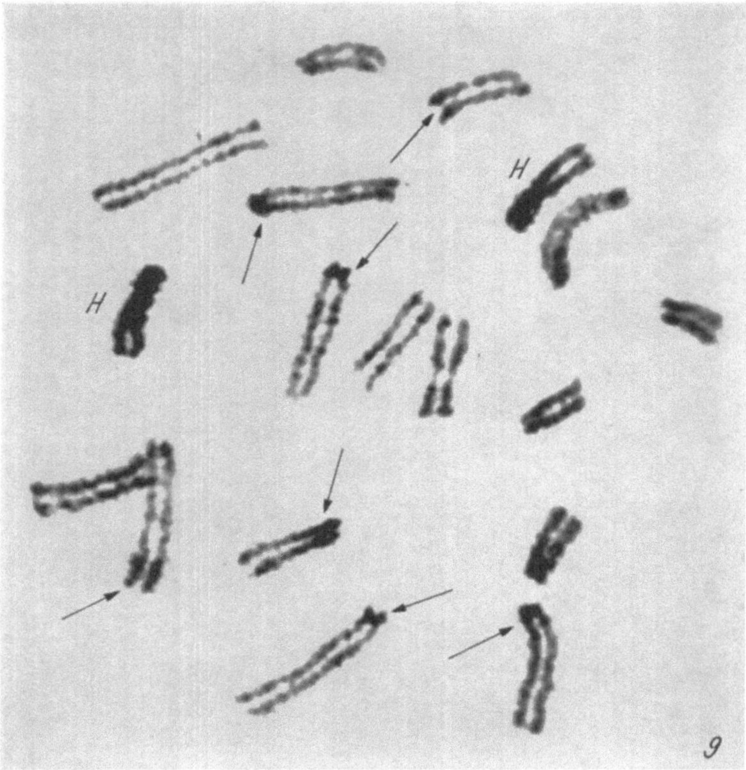

Fig. 9. Mitosis in the Bryophyte *Pellia epiphylla* (2n = 18) following 4 hours pre-treatment with 0.002 M 8-hydroxy-quinolene. Note two largely heterochromatic (H) chromosomes and the presence of short terminal heteropycnotic segments (arrows) in many of the others. Note also the absence of centric constrictions.

The notion of chromosome arms applies, of course, only in systems where centric activity is localised at one site. Two other types of centric organisation are known however. In one, mobility appears to be a property possessed by numerous loci scattered over a large region of the chromosome (polykinetic, see Fig. 8 *f*) while in the other it seems to be diffused over the whole chromosome (see pg. 126). This variation in centric organisation is never seen within complements or even within species. In fact, non-localised centric systems are confined to particular minority groups (JOHN and LEWIS 1965).

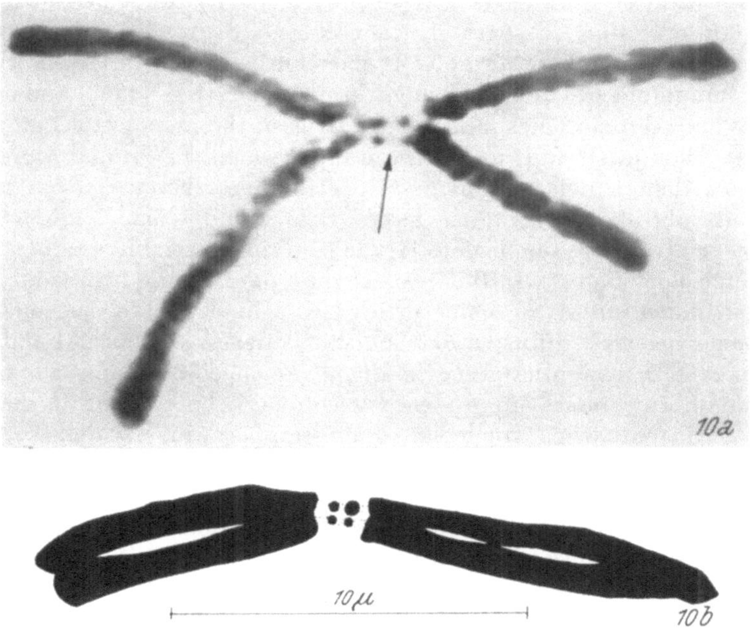

Fig. 10. The organisation of the kinetochore in *Allium cepa*; (Fig. 10*a*; photograph kindly supplied by Dr. Keith Jones) and *Hyacinthus orientalis* (Fig. 10*b*, after Lima-de-Faria 1956).

4. Secondary Constrictions

In addition to the primary, centric constrictions, chromosomes with localised kinetochores may also possess other regions which fail to spiralise and so form non-staining gaps. These are referred to as secondary constrictions. They may occur near the end of the chromosome so that the segment beyond the constriction is small. In such cases the term satellite (SAT) is used to describe it and the chromosome thread connecting the satellite to the main arm is referred to as the satellite stalk (Fig. 11 *a*).

The only known function of secondary constrictions is that of nucleolar organisation and in this they are commonly, though not invariably, involved. Thus chromosome 6 in maize is a nucleolar organising chromosome. The actual organising element is not, however, the satellite stalk itself but rather a prominent chromomere present at the base of the stalk—the nucleolar organising element. If this element is fractured experimentally by X-irradiation both the sub-units remain functional indicating that the organising element, like the centric organiser, is or can be a multiple structure (McClintock 1931).

A particularly clear example of the role of the secondary constriction has recently been provided in the clawed toad *Xenopus laevis*. Here a mutant heterozygote is known with only one nucleolus (1-nu) in each cell whereas wild-type tadpoles have basically two nucleoli in their diploid cells. The progeny obtained from mating two such (1-nu) heterozygotes fall into three groups having two (2-nu), one (1-nu) or zero (0-nu) nucleoli per cell. The ratio of these is 1 : 2 : 1 respectively as is expected of a typical mendelian mutation (ELSDALE, FISCHBERG, and SMITH 1958, FISCHBERG and WALLACE 1960). The heterozygotes (1-nu) can be shown to lack a secondary constriction in one of two homologous chromosomes (KAHN 1962) and doubtless both of these chromosomes lack constrictions in the homozygous anucleolate mutant. The latter fail to synthesise ribosomal RNA and possibly its precursors though it is normal in respect of non-ribosomal RNA fractions and in its ability to synthesise DNA (BROWN and GURDON 1964). In consequence, cells of the anucleolate larvae contain a variable number of loose, small nucleolar "blobs". In this respect they parallel the behaviour of those abnormal pollen grains in maize which lack a nucleolar chromosome.

In some species a diffusion of nucleolar material is a normal and regular occurrence. For example in *Oxalis dispar* no specific organiser is recognisable and at interphase up to ten nucleoli may be present (MARKS 1957). Notice also that when nucleolar organisers are inactivated in *Lolium*, nucleoli are then formed one at the end of each of the seven bivalents (JAIN 1957).

Filamentous structures have been demonstrated in the nucleoli of a number of plant species (ESTABLE and SOTELO 1951, 1955; LA COUR 1966). LA COUR (loc. cit) has given good evidence that these nucleolonemata are loops of chromosomal origin and represent the site of "nucleolar" DNA. Moreover he believes that these loops are directly concerned with the formation of the nucleolus. This view implies that nucleoli are products of specialised gene activity at a specific locus, the nucleolonemata arising as extensions of the chromosome axis comprising the nucleolar genes. In *Northoscordum inutile* four of the sixteen chromosomes are concerned with nucleolar organisation but they lack satellites and do not appear to possess secondary constrictions. Instead they are characterised by the possession of compound primary-cum-secondary constrictions which serve both kinetic and nucleolar-organising functions (KURITA 1964 and see Fig. 11) though, doubtless, these distinct functions are performed by different elements. Exaggerated primary constrictions of dual function have also been described in *Campanula persicifolia* (DARLINGTON and LA COUR 1959) and a comparable situation may obtain in *Poa alpina* (MÜNTZING 1948). In *Pisum*, on the other hand, it has been claimed that the nucleoli may be organised directly from the centromeres, a unique behaviour in which all the chromosomes take part but one which does not lead to over-extension of the constrictions (HÅKANSSON and LEVAN 1942, DARLINGTON and LA COUR 1950).

An especially interesting case of nucleolar organising chromosomes is found in the chicken. Microchromosomes occur rather widely in the *Amniota* though amongst mammals they are confined to the *Monotremata*. Because they occur in such profusion in the *Aves*, where they often approach the limit of resolution obtainable with a light microscope, there has been a

---

Fig. 11. Nucleolar organising chromosomes in *Tulbaghia dregeana* (Fig. 11*a*, photograph kindly supplied by Mr. C. G. VOSA), *Lilium pardalinum giganteum* (Fig. 11 *b*), and *Northoscordum inutile* (Fig. 11*c*, after KURITA 1964). Note heterozygosity for the nucleolar organising chromosome in 11*a* (arrow), terminal location of organisers in 11*b* (arrow), and compound nature of the constriction in 11*c*.

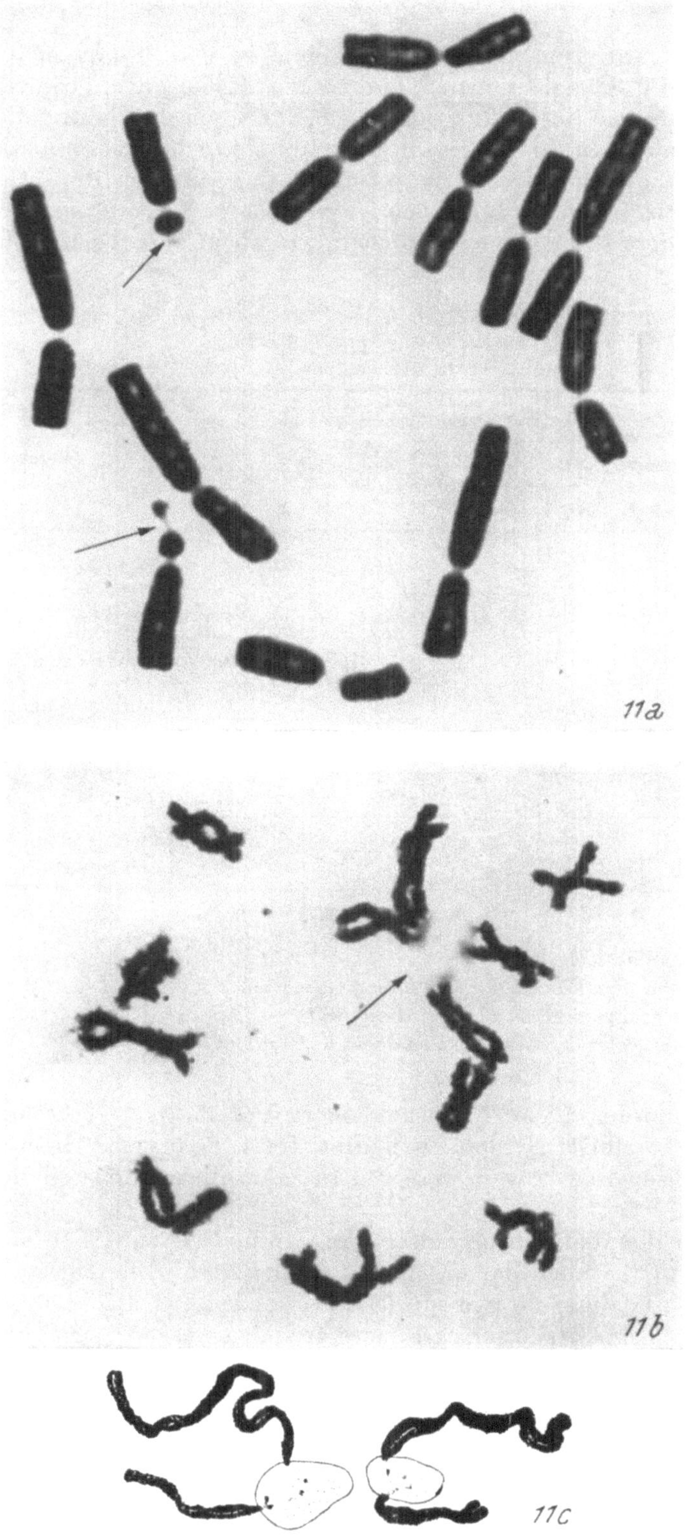
11a
11b
11c

tendency to regard them as insignificant or even non-chromosomal (NEWCOMER 1957, 1959). OHNO, CHRISTIAN, and STENIUS (1962) have, however, presented evidence that in *Gallus domesticus* the nucleolus is organised by microchromosomes. In this species it is possible to distinguish nine pairs of chromosomes, the fifth pair in order of decreasing size constituting the sex chromosome pair. Customarily, chromosomes from the seventh pair down are regarded as micro-chromosomes and there are about thirty pairs of them.

Table 4. *The Frequency and Distribution of Different Types of Nucleolar Organisers in Species of the Genus Trillium.* (After DYER 1964.)

Species	Distribution of Organisers											Reference
	A		B		C		D		E		Total	
	S	L	S	L	S	L	S	L	S	L		
1. *T. cernuum*	+										1	DYER 1964
2. *T. grandiflorum*	+					(+)		+	+		4	DYER 1964
3. *T. kamtschaticum*	(+)	(+)							+		3	MATSUURA 1938
4. *T. ovatum*			$+^{h}$						+		2	DARLINGTON and SHAW 1959
5. *T. recurvatum*					$+^{h}$						1	DARLINGTON and LA COUR 1940
6. *T. rivale*	+		+				(+)		$+^{l}$		4	DYER 1964
7. *T. sessile*			$+^{h}$			+	(+)		+		4	DARLINGTON and LA COUR 1940 DYER 1964
Total	4	1	3	—	1	2	2	1	5	—	19	

+ = Terminal nucleolar organiser
(+) = Nucleolar organiser on arm containing H-segment
$+^{h}$ = Nucleolar organiser attached to H-segment
$+^{l}$ = Long nucleolar organiser visible under all conditions

Nucleolar organisers cannot be located in any of the larger chromosomes but in each of thirty prophase figures from embryonic liver about twelve microchromosomes were observed in association with a single large nucleolus.

In *Trillium* the nucleolar organisers are terminal. DYER (1964) has made a comparison of the nucleolar organising properties of different species of the genus and finds that no two species have the same distribution pattern (Table 4). Nucleolar organisers are more frequent on short arms than on long ones (15 : 4) and they are most common on the short arm of the E-chromosome.

One final point of interest is that homologous chromosomes often differ in the size of their satellites. Heteromorphism involving the nucleolar organiser has been observed in plants of hybrid origin and following structural rearrangements involving the nucleolar constriction. The former condition depends upon the fact that the nucleolar organisers of related species may differ in strength and hence in competitive ability. When these species are crossed the stronger organisers take over the entire nucleolar activity of the cell. In consequence the metaphase morphology of the chromosome containing the non-functional organiser is modified since no secondary constriction develops in it. For example *Crepis alpina, dioscoridis, neglecta*, and *tectorum* lose their constrictions in competition with that of *C. capillaris* while that of *capillaris* is itself suppressed by *C. parviflora* (Fig. 12).

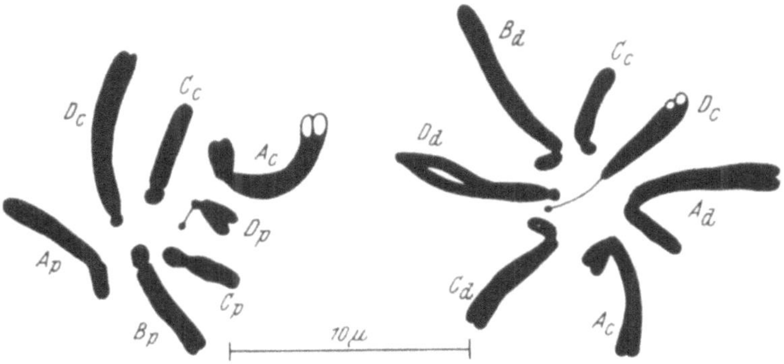

Fig. 12. Competitive suppression of secondary constrictions in two species hybrids of *Crepis*, *capillaris* × *parviflora* (left) and *capillaris* × *dioscoridis* (right). The capital letters signify the members of the sets while the subscripts refer to the species from which they come. Note that while the *capillaris* organiser is suppressed in the c × p hybrid it is functional in the c × d (after NAVASHIN 1934).

The net result of this is that the mitotic complements of hybrids and allopolyploids may fail to reveal all the organisers present in the parental sets. What may well prove to be a functionally comparable case appears to exist in *Cavia cobaya*. Here the largest pair of chromosomes in the complement is invariably heteromorphic at mitosis; one member only of this acrocentric pair has a distinct secondary constriction forming a SAT-zone in the middle of its short arm. This heteromorphism persists into meiosis and in all the prophase nuclei examined the larger chromosome was invariably associated with a nucleolus (OHNO, WEILER, and STENIUS 1961). The non-participating short arm, on the other hand, was intensively heteropycnotic. OHNO *et al.* are of the opinion that this heteromorphism does not imply any inherent structural difference between the homologues. Rather, they suggest that both homologues have an inherent capacity for nucleolar organisation but only one actually participates in this function (but see pg. 88). Odd numbers of nucleolar organising chromosomes are known also among the autosomes in *Ambystoma tigrinum* (DEARING 1934) and *Rattus norvegicus* (OHNO and KINOSITA 1955) and the X-chromosome pair of female *Cricetus griseus* (YERGANIAN *et al.* 1960). These instances are important because they suggest that other homologous regions of a chromosome pair within the same nucleus need not function synchronously (see also pg. 12). In this connection HAUSCHKA and BRUNST (1964) have described a unique sex-difference in the nucleolar autosome of *Siredon mexicanum*.

This chromosome ranks third in size in the male whereas in the female it is about the sixth largest. This size difference is due to a significant length dicrepancy in the short arm which carries the nucleolar organiser (Table 5). The authors suggest that the sex-difference in the nucleolar arm ratios may be viewed as a hormonally conditioned difference in the timing of nucleolar synthesis during interphase and nucleolar dissolution during early prophase which differentially affects the degree of contraction in the nucleolar arms during metaphase. We appear to be dealing here with a secondary sex influence on chromosome function and morphology.

Table 5. *Percentage of Total Karyotype Length Taken up by the Short and Long Arms of the Satellited Nucleolar Chromosome in Male and Female Siredon mexicanum.* (Data of Hauschka and Brunst 1965.)

Short Arm (nucleolar organising)		Long Arm	
♂	♀	♂	♀
$\bar{x}_{10} = 4.32 \pm 0.16$	$\bar{x}_{10} = 3.41 \pm 0.20$	$\bar{x}_{10} = 5.74 \pm 0.12$	$\bar{x}_{10} = 5.60 \pm 0.11$
Sex difference $= 0.91 \pm 0.26$ $P < 0.01$*		Sex difference $= 0.14 \pm 0.16$ $P > 0.20$ Not sig.	

5. Telomeres

Chromosome ends have special properties several of which resemble those of centromeres (John and Lewis 1965). Some of their peculiar features may simply be a reflection of their terminal position but it has been argued that telomeres have a special structure which confers a polarity upon them. Thus Muller (1940) conceived of the telomere as a region terminated by a special telogene which may have been compound and complex and which conferred the properties of polarity. This polarity, amongst other things, prohibited union with similar ends or with fracture sites produced by breakage so that telomeres could not come to occupy interstitial positions. Conversely it has been argued that, since they lack the special telogenic organisation, new ends produced by breakage could not function as stable ends. There is, however, some evidence for the stabilisation of ends of both centric and interstitial origin (see pg. 147) and some support for the idea that original ends may become labile especially after mutagenic treatment (Darlington and Koller 1947, pg. 199).

In view of the properties shared by the ends of even non-homologous chromosomes a similarity of structure is to be expected and is found (Lima-de-Faria and Sarvella 1958 and see Fig. 13). White (1957, 1961), however, on the basis of end pairing in complements where linear pairing is impeded by structural rearrangements, has argued that the ends of many, if not all, the chromosomes in a complement may be homologous. This homology may be confined to "minute terminal regions (a telomere plus perhaps 1–3 genetic loci)." Clearly, therefore, White has in mind a much shorter region than that usually regarded as constituting the telomere region. Thus Lima-de-Faria and Sarvella (*loc. cit.*) describe the telomere as a "compound structure consisting of several differentiated segments and not a single band or chromomere." In rye, for example, they regard this region as composed of at least eight segments—four chromomeres and four fibrils. Of course, while the end of the end exists, the beginning of the end is a position that needs to be defined. To some extent LIMA-DE-FARIA and SARVELLA use a functional criterion and argue that all the regions which compose the telomere possess the essential property of monopolarity so that breakage within the telomere will still produce stable ends. Evidence for this polarity is derived from the production of stable terminal deficiencies in rye, maize and elsewhere which involve breakage in various segments of the telomere.

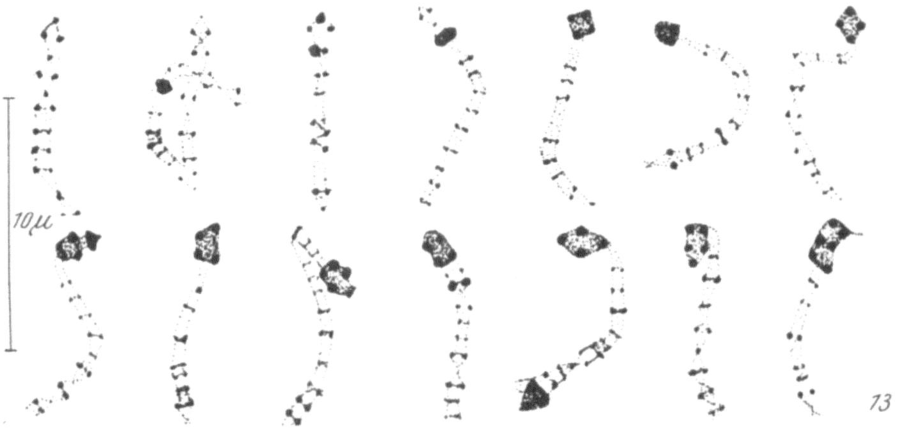

Fig. 13. The 14 telomeres of the 7 bivalents in a single cell of rye (*Secale cereale*, 2n = 14) as seen at pachytene (after LIMA-DE-FARIA and SARVELLA 1958).

On this basis the telomere shares a compound organisation with both the centromere and the nucleolar organiser. Of course, the production of stable telocentrics following misdivision (see pg. 112) is not be expected unless a comparable polarity exists in, or can be acquired by, the centromere. Telomere and centromere must be co-extensive in telocentrics. Interestingly, as we have seen, centric and nucleolar functions are co-located in some cases while terminal nucleolar organisation has frequently been described. What is more, structurally unaltered ends may exhibit neo-centric activity (see JOHN and LEWIS 1965).

II. The Genetic Continuity of the Karyotype

It should now be clear that the phenotype of the chromosome is not constant. It varies not only during the division cycle but between cycles. Thus the chromomeric organisation is evident only during the early stages of division, allocycly may be differential not only between segments but between cell types. Thus in many of the *Acrididae*, the *Gryllidae* and the *Tetrigidae* the change from negative to positive heteropycnosis of the single

X-chromosome of the male germ line is progressive. Again the length of the chromosomes varies not only at different stages of a particular cell cycle but even at the same stage of different cell cycles. For example in *Papaver rhoeas* the metaphase chromosomes from leaf meristem mitoses are about half the length and half the thickness of the equivalent stage in the root meristem (Fig. 14*a* and *b*). Likewise LEVAN (1933) claims that in *Allium fistulosum* the metaphase chromosomes are shorter in the pollen than in the root tips (Fig. 14*c* and *d*) while BERRIE (1959) finds that in the cycad *Encephalatos barteri* the nine chromosomes of the haploid complement are smaller at the second pollen grain mitosis than at the first. Equivalent situations exist in animals too. In the early embryonic stage of grasshoppers

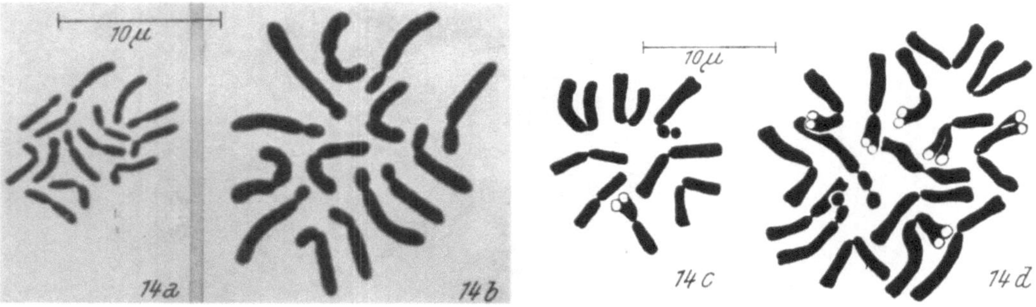

Fig. 14. Variation in size of somatic chromosomes in different cell types of the same species. Figs. 14*a* and *b*, diploid metaphases from the leaf (14*a*) and root (14*b*) meristem of *Papaver rhoeas* (after HASITSCHKA-JENSCHKE 1956). Figs. 14*c* and *d*, haploid pollen grain (14*c*) and diploid root (14*d*) mitosis of *Allium fistulosum* (after LEVAN 1933).

and locusts a fixed number of very large neuroblasts are present all of which undergo repeated and unequal cell division. Of the products of this unequal division only the larger daughter retains the morphological and physiological characteristics of the neuroblast. Not only are neuroblast cells of large size, so are the chromosomes they contain (Fig. 15). Likewise, as the SCHRADERS point out, the chromosomes of pentatomids are distinctly larger in the early generations of spermatogonial cells than those of later generations (SCHRADER and HUGHES-SCHRADER 1956). Finally whereas in fibroblast-like cells from the testis of *Cricetulus migratorius* the sex chromosomes are the second largest in the complement in lung and spleen they are consistently the third largest (YERGANIAN and PAPOYAN 1965). An equivalent situation exists also in polytene systems. In *Chironomus tentans*, for example, the fully developed polytene chromosomes of the salivary gland are about ten times longer than the chromosomes at pachytene. It has been found that the length of the salivary gland chromosome is mainly a function of lateral multiplicity which is, of course, affected not only by polytenisation but by pairing too. Each duplication increases length by 15–20% in the salivary gland chromosomes. Nevertheless, even though the chromosomes of the malpighian tubules and mid gut undergo four or five fewer replications than those in the salivary glands they are up to one and a half times longer (BEERMANN 1952).

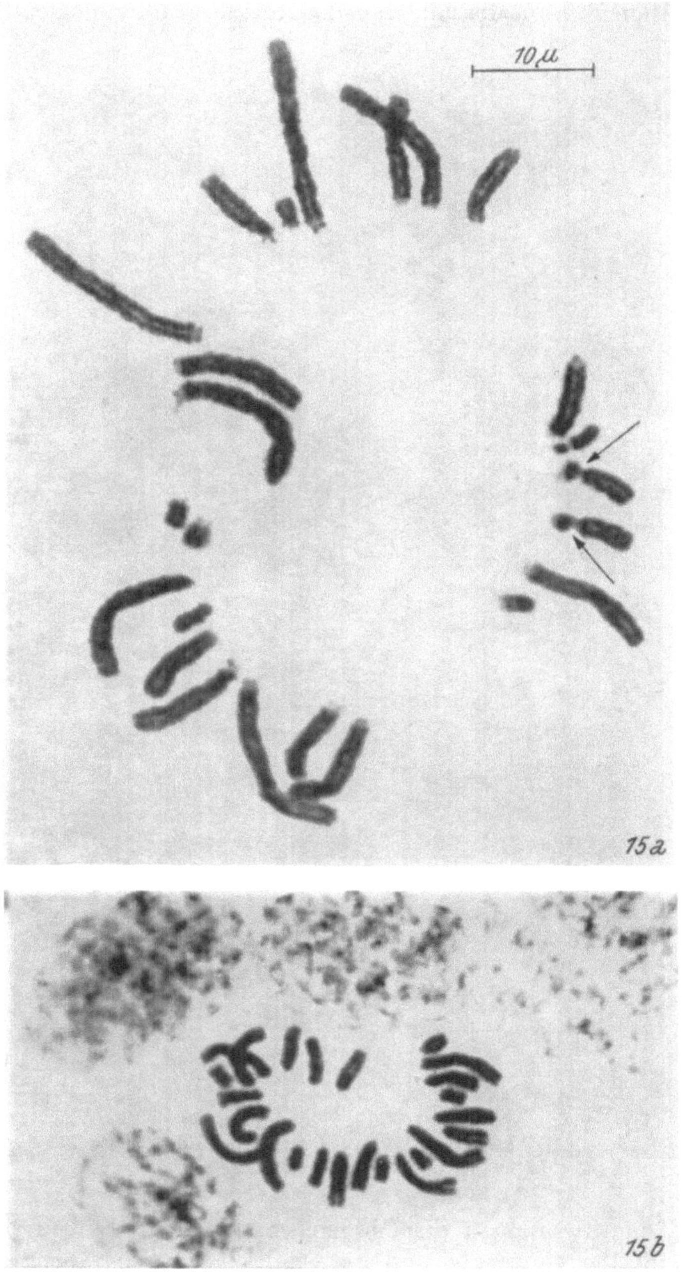

Fig. 15. Mitotic metaphases in the two distinguishable cell types found in the embryo of *Schistocerca gregaria*. The upper plate (15*a*) is taken from a neuroblast and is shown at the same magnification as the lower (15*b*). Note that the secondary constrictions present in 15*a* are not developed in the smaller cell type. Both types, however, have the same DNA content.

Despite variations of this sort, the sequences of kinetochore, heterochromatic regions, nucleolar organisers and other secondary constrictions are usually sufficiently specific and constant to confer a cytologically-recognis-

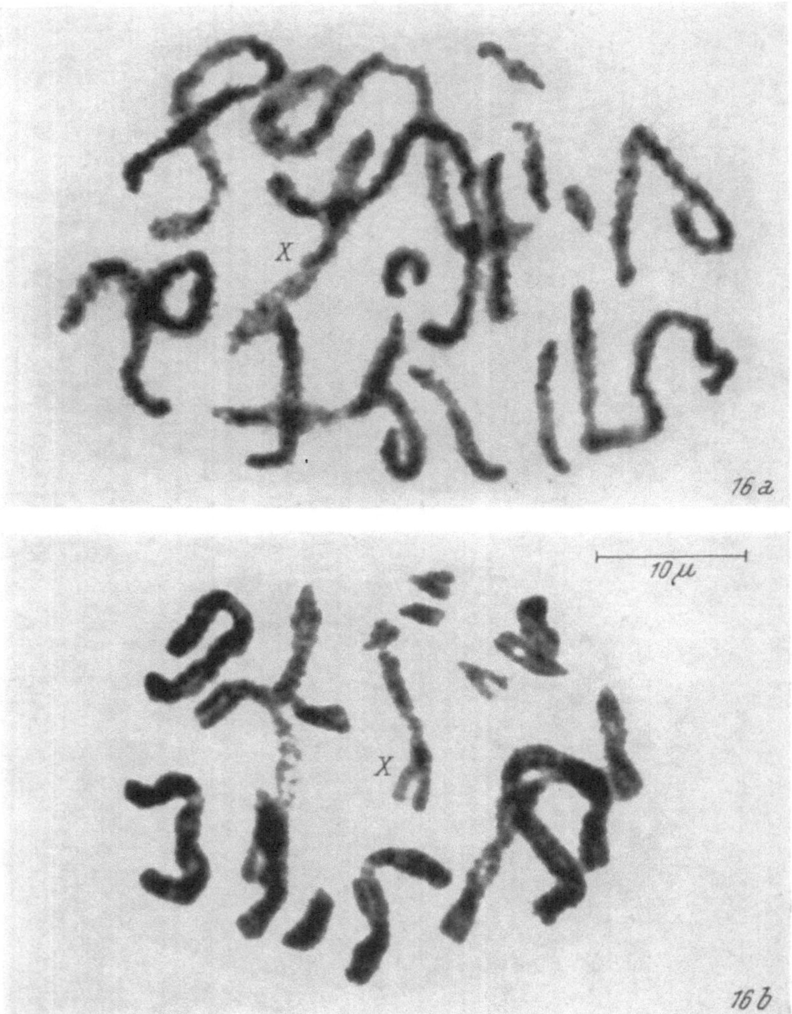

Fig. 16. Chromosome polarisation in early (Fig. 16*a*) and late (Fig. 16*b*) prophase cells of the grasshopper *Myrmeleotettix maculatus*.

able individuality upon a chromosome complement. Indeed the essential constancy of the karyotype and its usually accurate transmission during cell division has been recognised since the time of the early cytologists. RABL (1885), for instance, provided conclusive evidence that the chromosomes maintained not only the orientation (Fig. 16) which they displayed at the close of division but also their identity. In consequence when chromosomes reappear at prophase in rapidly dividing cells they agree in position, in number and in structure with those that entered into the formation of the resting nucleus. Every chromosome that appears in a nucleus thus has some kind of direct continuity through successive cell divisions and this continuity is important in allowing the karyotype to function as the physical basis of the genotype.

Two series of observations support this contention of direct continuity. At telophase, relic spirals are visible in each chromosome, a consequence of uncoiling. Equivalent relic coils are evident also at the ensuing prophase. In this respect prophase must be regarded as a continuation of telophase rather than a reversal of it for the cycle of uncoiling, though temporarily in abeyance during interphase, is resumed at the onset of the next mitotic cycle. The mechanical relationship and continuity between the relic coils of successive division sequences clearly implies a permanence of essential chromosome organisation but need not, of course, imply an actual retention of the linear structure of the chromosome throughout the whole of interphase. Three further lines of evidence suggest, however, that this is in fact so:

(a) Parts of chromosomes can sometimes be shown to reappear in prophase in just those regions of the nucleus where they disappear at telophase.

(b) Occasionally one or more chromosomes remain inside distinct vesicles or karyomeres or else persist as heteropycnotic bodies throughout interphase, and

(c) In first generation hybrids between species with clearly distinguishable complements each of the chromosome types which characterise each parent are represented in the hybrid karyotype (see pg. 106 and Fig. 12).

Every chromosome that appears at mitotic prophase thus corresponds with a chromatid that entered into the formation of the resting nucleus at the preceding telophase in all its visible properties except its doubleness. The one consistent genetic change that the chromosome undergoes during interphase is that of duplication. This involves both the synthesis of new molecular constituents (replication) and their organisation into a new chromatid (integration or individuation). The process can be studied indirectly by allowing the chromosomes to replicate in the presence of a suitable labelled precursor, most commonly tritiated thymidine, whose distribution pattern in the chromosome can then be followed in terms of the effects of the radioactive components on a photographic emulsion. Using this technique it has been repeatedly found that although both chromatids are fully labelled at the first post-treatment metaphase (X_1) there is a label segregation at the next (X_2) metaphase, each chromosome showing one labelled and one unlabelled chromatid. This demonstrates most effectively that the process of replication must be semi-conservative in character for the actual molecules which comprise the chromosome prior to the uptake of the labelled elements are still present at the second mitotic cycle following duplication.

The chromosomes, though potentially permanent, are also inherently subject to change in the proportions, positions and properties of their component organelles so leading to differences in number, structure, size and behaviour. It is to such changes that we must now turn but before doing so we would draw the attention of the reader to a most significant fact. Chromosome size is subject to genotypic as well as environmental control; its extent can therefore be altered by changes at the genic level following either mutation or recombination. The position of the centromere, and thus the shape of the chromosome is not so easily altered. To effect such a change the chromosome must be broken and reorganised structurally. Therefore while the limits set by the genotype can often be extended by recombination the structure of the chromosome imposes a far more rigid restriction. Further, while structural changes will clearly affect only the chromosomes involved in breakage and reunion, genotypically-determined changes will have more general effects on the karyotype.

III. Karyotypic Variation During Development

1. Genetic Mosaicism

The accurate perpetuation of the karyotype, and thus the genotype, is the rule during both development and heredity. However, chromosome changes of a genetic kind do occur during development. Some of these are regular features of normal development in particular organisms so that their incidence and time of occurrence can be predicted with extreme accuracy. Indeed there is evidence that the normality of development is disturbed unless these anomalous events obtain. In other cases the mutations are not integral parts of development and their existence is less predictable. Even so particular kinds of developmental variation may be associated with particular organisms or chromosomes so that their existence can be anticipated to some extent. In this sense, therefore, they are rather different from the sporadic and unpredictable mutations which are liable to occur at any stage and in any organism.

a) Regular

i) Somatic Mosaicism

Undoubtedly the commonest genetic change to which somatic nuclei are subject is the apparently uniform multiplication of the genetic material leading to the production of nuclei with 2, 4, 8, and so on, times as much DNA as the original nucleus. All such cases depend upon the occurrence of two or more successive replication cycles during interphase. This process has been referred to as an endoreduplication (LEVAN and HAUSCHKA 1953) or an endomitosis (GEITLER 1939). The term endomitosis was first applied to cases which included detectable signs of the process during interphase whereas endoreduplication gave no such evidence. Both agree, however, in that no spindle is formed after replication is complete and we shall use the one term endomitosis to cover all the variants. For the most part cells affected by this process do not subsequently enter division. Consequently the exact effect on the karyotype cannot be observed. In some cases the process takes place while the chromosomes are in a condition of "permanent prophase" while in others the affected cells can be brought into division by hormone treatment or other artificial means.

From a study of such cells it is clear that endomitosis can take two main forms (Fig. 17). Thus, in tissues like the salivary glands of *Diptera*, the products of DNA replication are not individualised so that the number of distinct chromosomes is not increased. Rather it is the degree of lateral

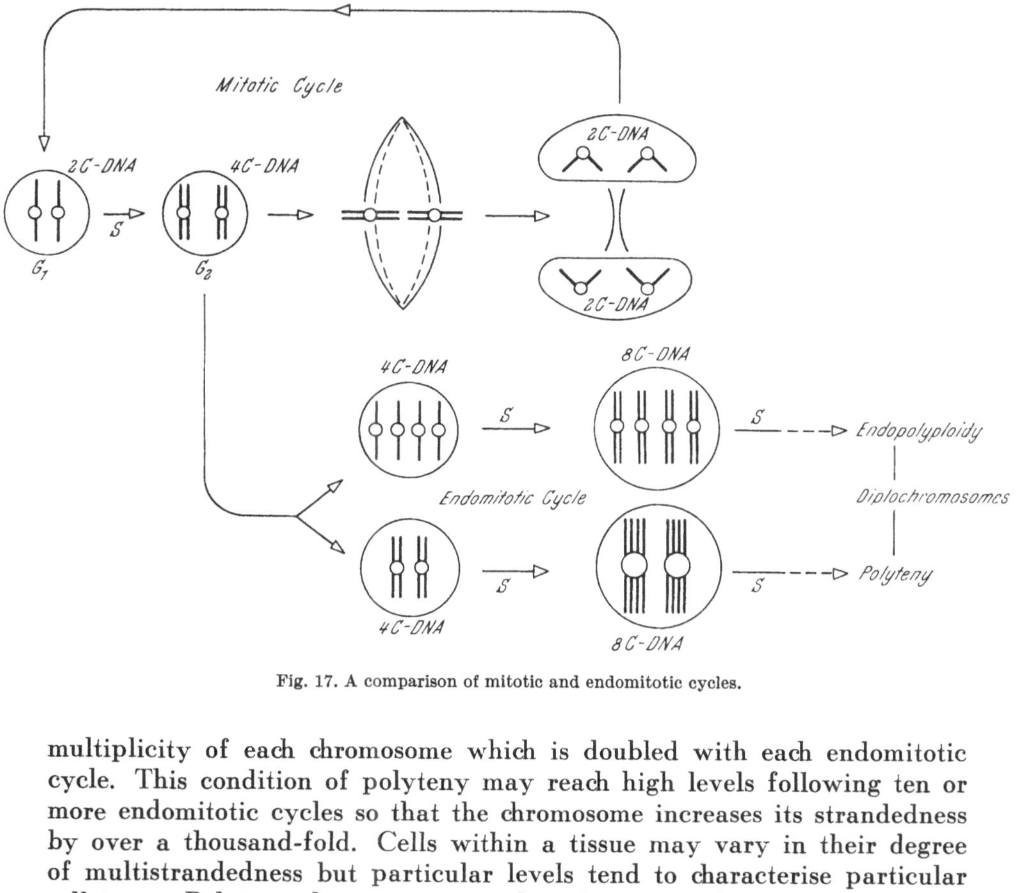

Fig. 17. A comparison of mitotic and endomitotic cycles.

multiplicity of each chromosome which is doubled with each endomitotic cycle. This condition of polyteny may reach high levels following ten or more endomitotic cycles so that the chromosome increases its strandedness by over a thousand-fold. Cells within a tissue may vary in their degree of multistrandedness but particular levels tend to characterise particular cell types. Polytene chromosomes are best known in the *Diptera* (Fig. 18) but they have recently been described in the macronucleus of the ciliate *Stylonychia muscorum* (Alonso and Pérez-Silva 1965). They are known also in plants under normal (Fig. 19, Table 6) and abnormal (Fig. 20) conditions. There are, however, undoubted differences between the nature of the polytene chromosomes found in the *Diptera* and those recorded elsewhere. Thus in plants they are never banded while in *Stylonychia* according to Ammermann (1965) they represent a transient stage after which, curiously, the nucleus loses much of its DNA. This suggests that they are polytene chromosomes of a very specialised type.

Alternatively, endomitosis may involve changes in the constitution of chromatids so that the number of distinct units is increased geometrically. Some indication of this process can sometimes be observed during the endomitotic sequence itself. Its consequences are best seen, however, when the affected cells are brought into division. At such a division it can be seen that chromosomes which have undergone a single extra duplication consist

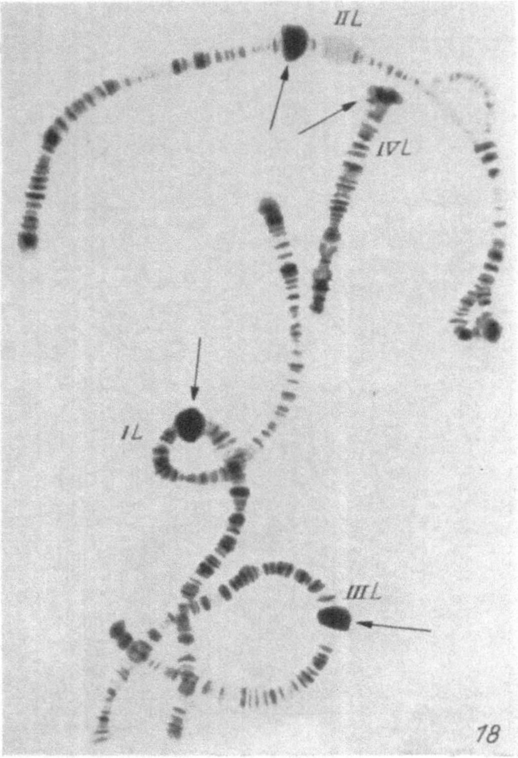

Fig. 18. Polytene chromosomes from the salivary glands of the midge *Glyptotendipes barbipes* (2n = 8). The complement consists of 3 pairs of long metacentrics and one pair of short telocentrics, which are numbered I to IV in decreasing order of size. The kinetochores (arrows) are marked by prominent heterochromatic segments (after BASRUR 1957).

not of two chromatids but of four attached as a unit. Each of these "diplochromosomes" units form two normal chromosomes by metaphase which thus contribute at ana-telophase to the formation of tetraploid daughter nuclei. Nuclei with 16 C-DNA values show 8-parted chromosomes but higher DNA values are rare except in nutritive tissues associated with embryos.

Somatic polyploidy is well documented in the vegetative development of flowering plants where it leads to the production of mixed cell populations (Table 7). For example, while the basic complement is retained in the apical meristem, the cambium, pericycle and germ-line, polyploidisation commonly occurs in the cortex and pith, and the nuclei of the xylem vessels become highly polyploid prior to their degeneration (Fig. 21). Poly-

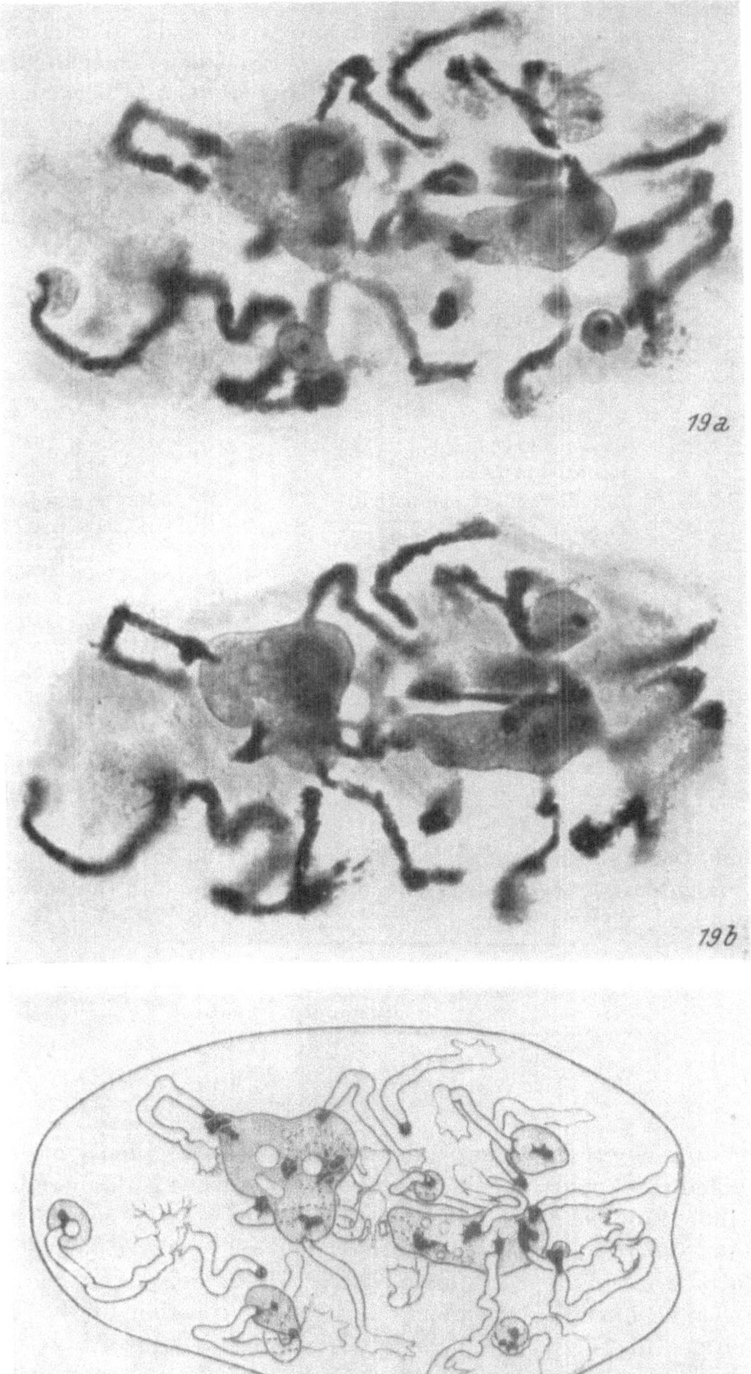

Fig. 19. A giant 192-ploid nucleus from the haustorium at the chalazal end of the endosperm of *Rhinanthus alectorolophus* to show the development of polytene chromosomes (after TSCHERMAK-WOESS 1957).

ploidisation is known also in fruits, in nutritive tissues like tapetum and endosperm and in glandular hairs. The synergid nuclei associated with the egg cell in the embryo sacs of flowering plants frequently become enlarged. In *Allium angulosum* and *A. pulchellum* this is associated with endoreplication to give octoploid nuclei. A similar process is known also in the antipodal nuclei of these and other species (Table 8). Typically, the endosperm of angiosperms is originally triploid. Endopolyploidisation may, however, increase this level considerably. Particularly high levels of endo-

Table 6. *Polytene Chromosomes in the Ovules of Plants.*

Tissue	Species	Reference
1. Antipodal	*Aconitum*	TSCHERMAK-WOESS 1956
	Allium ammophilum	ERBRICH 1965
	Hordeum jubatum (following rye pollination)	BRINK and COOPER 1943
	Papaver rhoeas	HASITSCHKA 1956
2. Endosperm	*Allium ursinum*	GEITLER 1955
	Brassica oleracea (after autotetraploid pollination)	HAKANSSON 1956
	Zea mays	DUNCAN and ROSS 1950
3. Haustoria	*Thesium alpinum* and *linophyllon*	ERBRICH 1965
4. Suspensor	*Phaseolus coccineus*	NAGL 1965
5. Synergid	*Allium nutans* and *odorum*	HAKANSSON 1957

polyploidy have been reported by STEFFEN (1956) in the endosperm of *Pedicularis palustris* where different levels are attained in the various zones as follows:

Zone	Region	
	Micropylar	Chalazal
Non-haustorial	6 x	12 x
Haustorial	192–384 x	96 x

TSCHERMAK-WOESS (1956) has estimated that endopolyploidy occurs in the non-lignified root tissues of some 140 species of flowering plants representing twenty-four families. On the other hand thirty-nine species, including *Helianthus tuberosus* show little or no tendency towards polyploidisation either *in vivo* or in tissue culture. Clearly, therefore, the existence of this genetic change during development is itself a reflection of the genotype. Combinations of polyteny and endopolyploidy are known also and there are indications that the former can lead to the latter (BAUER 1938, WHITE 1948, BIER 1957, and MATUSZEWSKI 1965).

In the majority of flowering plants only one of the three meiotic products is functional on the female side. Typically it divides three times to

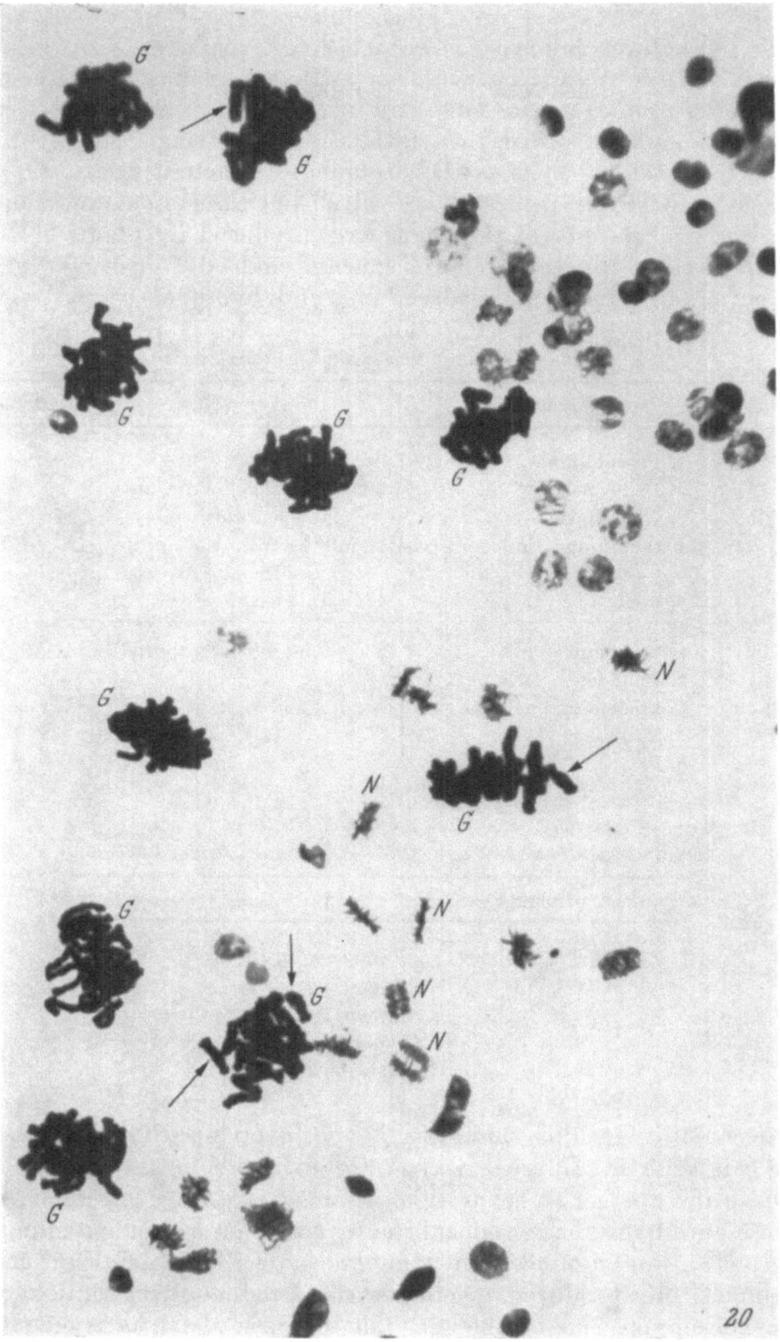

Fig. 20. The chalazal region of the endosperm of *Tulbaghia dregeana* × *T. violacea* to show the development of giant polytene nuclei. Normally this tissue divides only twice to give 4 large nuclei which remain in interphase. In this one endosperm, however, there were 16 giant nuclei (G) of which 10 are shown in the photograph, and all of them have come into mitosis. Each chromosome within these giant nuclei is polytene and is itself larger (arrows) then the whole ordinary diploid complement (N) which can also be seen in mitosis (Photograph kindly supplied by Mr. C. G. Vosa).

give an 8-nucleate embryo sac in which all the nuclei are genetically identical. In some cases, however, two (*Allium* and *Endymion*) or all four (*Adoxa, Peperomia*) megaspores are involved in producing a common female gametophyte. Clearly, in these cases, the female haploid phase is invariably mosaic so long as the diplophase is heterozygous. Of course, only genic differences are usually involved but chromosome mosaicism can occur. In fact, a study of the endosperm produced by plants of *Trillium grandiflorum* heterozygous for H-segments enabled Rutishauser (1957) to carry out a half-tetrad analysis (see Lewis and John 1963).

Table 7. *Somatic Polyploidy in Plant Tissues.*

Region	Tissue	Level of ploidy	Reference
Onion Root	1. Procambium	2 x	D'Amato and Avanzi 1948
	2. Pericycle	2 x with isolated 4 x cells	
	3. Epidermis	2 x and 4 x (mitosis rare)	
	4. Cortex and Endodermis	Mainly 4 x mitosis, some 2 x or exceptionally 8 x	
	5. Stele	4 x with 8 x xylem cells	
Bean Stem	1. Cambium	2 x	Dermen 1941
	2. Endoderm 3. Phloem parenchyma 4. Medullary rays 5. Pith	4 x, 8 x and 16 x	

Table 8. *Polyploid Antipodal Nuclei in Plants.*
(Data of Tschermak-Woess 1957, Hasitschka-Jenschke 1959.)

Level of ploidy	Species
8 x	*Allium species, Helleborus niger*
16 x	*Othonna crassifolia*
32 x	*Anemone hepatica, Clivia miniata*
64 x	*Eranthis hiemalis, Kleinia ficoides*
128 x	*Aconitum, Papaver*

Superimposed on this mosaicism is one involving differences in ploidy which can arise in different ways. For example in the classic case of *Fritillaria* there is a 1 : 3 separation of nuclei at the 4-nucleate stage. The chromosomes of the three chalazal nuclei orient on a common spindle and their division thus produces, not 6 haploid nuclei but 2 triploid ones. One of the nuclei thus produced, together with a product of the haploid nucleus at the micropylar end of the sac, join to form the primary endosperm nucleus which is thus 4 x and the endosperm itself becomes pentaploid.

ii) Gonadosomic Mosaicism

As a rule the chromosomes of the somatic mitoses agree closely with those found in the germ line (Fig. 22). There are, however, cases where a regular and predictable difference in chromosome number exists between soma and germ line. The classical example of this is *Parascaris equorum*. Here the situation is complicated by the fact that the germ-line chromosomes (Fig. 8*f*) are compound and polycentric in character and they dissociate into a much larger number of simple monocentric elements during somatic development. Let us therefore turn to those cases where no such complication occurs.

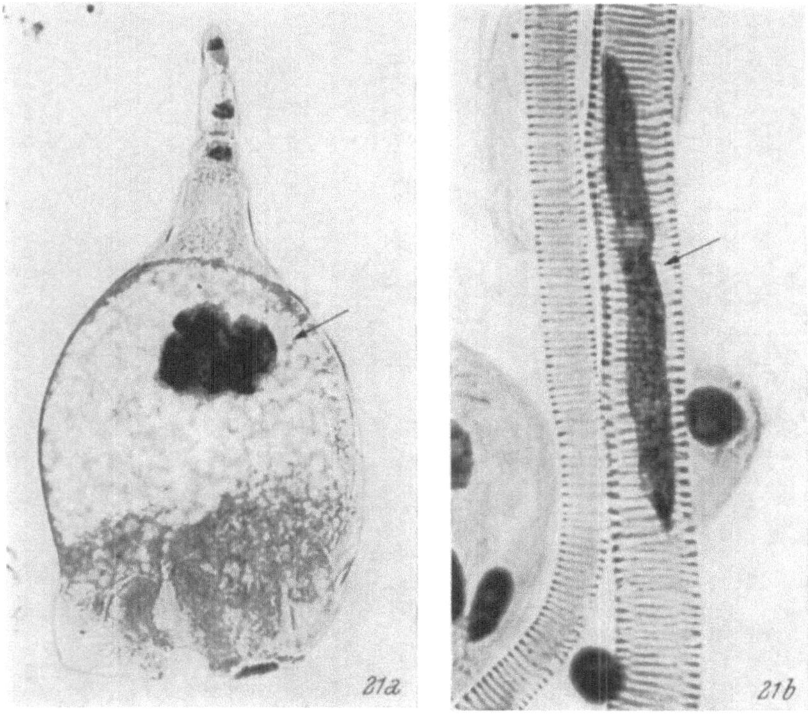

Fig. 21. Endopolyploid nuclei (arrows) in a basal cell of an epidermal hair from the anther of *Bryonia dioica* (Fig. 21*a*) and a developing protoxylem vessel from the stem of *Vicia faba* (Fig. 21*b*).

In three dipteran groups—the *Cecidomyidae*, the *Sciaridae* and the *Orthocladiinae* (*Chironomidae*)—there are anomalous chromosome cycles leading to the differential elimination of chromosomes from somatic nuclei during the early cleavage mitoses. In the orthoclads (Table 9) and the cecidomyids (Table 10) the chromosomes eliminated are called E-chromosomes and their number varies from 2–52 in the former and from 8–67 in the latter. In the *Sciaridae* on the other hand elimination involves the paternal X-chromosome and, where they are present, the so-called "limited" or L-chromosomes. In the cecidomyids and the sciarids the situation is complicated by two further factors. First, different patterns of elimination may be found in the male as opposed to the female soma. Second, the anomalous somatic cycle is associated with modifications of the meiotic system which, like the somatic variation, is also differential between the

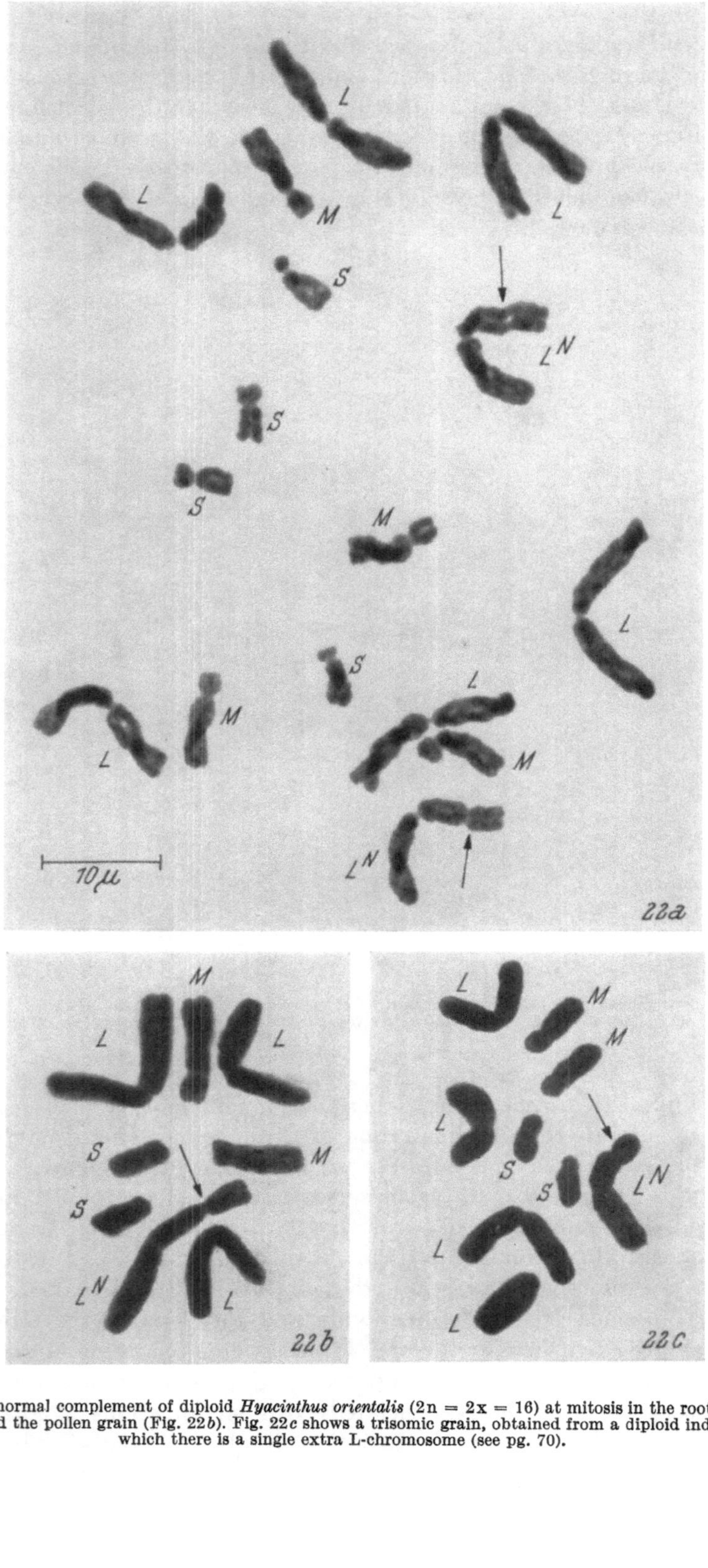

Fig. 22. The normal complement of diploid *Hyacinthus orientalis* (2n = 2x = 16) at mitosis in the root meristem (Fig. 22*a*) and the pollen grain (Fig. 22*b*). Fig. 22*c* shows a trisomic grain, obtained from a diploid individual, in which there is a single extra L-chromosome (see pg. 70).

sexes. The net result is a process of complementary gametic elimination (Table 11).

Two further instances are known in animals where, although no somatic mosaicism occurs, there is a differential behaviour in the germ line. The first of these is found in *Enchytraeus lacteus*. Among the polyploid cytotypes known to occur in this species (2 n = 18, 36, 54, 72, and 170) the 170 chromosome form includes 8 chromosomes which are considerably larger than the remaining 162. During gametogenesis the former give rise to four bivalents

Table 9. *S and E Chromosome Numbers in the Orthocladiinea.* (Data of Bauer and Beermann 1952.)

Species	No. of Chromosomes	
	S-type (soma and germ line)	E-type (germ line limited)
Metriocnemus cavicola	4	24–52
inopinatus	4	12–16
hygropetricus	6	2–8
spec.	6	4–6
Psectrocladius obvius	6	2–8
platypus	6	6–10
remotus	6	20–28
spec.	6	ca. 12
Trichocladius vitripennis	6	10–14
Clunio marinus	6	ca. 16
Limnophyes spec.	6	ca. 16
Eucricotopus atritarsis	6	ca. 18
silvestris	6	20–24
Acricotopus lucidus	6	34–38

while the latter occur as only univalents. In the female the four large bivalents behave normally at the first oocyte division and since the univalents divide only at the second division the female pronucleus contains 4 L and 162 S chromosomes. In the male meiosis, however, the 162 small chromosomes are eliminated at first division so that the male pronucleus transmits only the four L-chromosomes (Christensen and Jennsen 1964).

Planarian worms are characterised by free and undifferentiated neoblast cells which remain quiescent during the embryonic period but subsequently give rise to the germ cells. In *Dugesia lugubris* and *benazzii* and in *Polycelis nigra* and *tenuis* there exist amphigonic diploid biotypes with a normal chromosome cycle and pseudogamic polyploid biotypes where chromosome behaviour is anomalous. For instance in *D. lugubris* (Benazzi 1966) the polyploid (3 x and 4 x) biotypes fall into two groups according to whether the oogenesis is meiotic or ameiotic in character. In both cases the male germ line is diploid for sexual differentiation involves the elimination of one (3 x) or two (4 x) haploid sets from the polyploid neoblasts. Following this the male line remains normal in behaviour although the function of

Table 10. *Chromosome Numbers in the Cecidomyidae.* (After Camenzind 1966.)

Sub-family	Species	Female			Male			Reference
		Germ Line	Soma	Meiosis-I	Germ Line	Soma	Sperm	
Heteropezinae	*Heteropeza pygmaea*	77 66	10 10	 33 II	58 (68) 53 (63)	5 5	7 7	Hauschteck 1962 Camenzind 1966
	Miastor metraloas	48	12		48?	6	6	Kraczkiewicz 1937, White 1946
	Miastor spec.	39 (40, 38)	8					Nicklas 1959
Lestremiinae	*Mycophila speyeri*	29	6	14 II + 1 I 13 II + 3 I	29?	3	3	Nicklas 1960
Cecidomyiinae	*Asphondylia monacha*		8		over 50		over 4	White 1950
	Lasioptera asterspinosae		8				4	White 1950
	Lasioptera rubi	ca. 40	8		ca. 40	6	4	Kraczkiewicz 1950
	Mayetiola destructor	ca. 40	8			8		Bantock 1961
	Monarthropalpus buxi	ca. 50	8	4 II + ca. 42 I	ca. 50	6	28	White 1950
	Mikiola fagi	24	8	4 II + 16 I				Matuszewski 1962
	Oligotrophus pattersoni	34	8	4 II + 26 I	ca. 34	6	10	White 1950
	Phytophaga celttiphyllia	ca. 36	12	6 II + ca. 24 I				White 1950
	Rhabdophaga batatas		8		40 (38, 36)		4	Geyer-Duszyńska 1961
	Rhopalomyia sabinae	ca. 25	8	4 II + ca. 17 I	25	6	5, 7, 8	White 1950
	Taxomyia taxi	ca. 40	8		40	6	4	White 1947
	Trishormomyia helianthi	24	8	4 II + 16 I	24?	6	4	White 1950
	Wachtiella persicariae	ca. 40	8		ca. 40	6		Geyer-Duszynska 1959

the haploid sperm is limited to activation of the egg. In the ameiotic females the germ line retains the ploidy level of its parental neoblast. In meiotic females, however, there is a duplication of the chromosome complement at the time of the differentiation of the neoblasts. Thus in triploid individuals the oogonial cells become hexaploid and these go through a regular meiosis to give triploid pseudogamic egg cells.

Finally, a unique elimination process occurs in the creeping vole, *Microtus oregoni*. Here male zygotes are XY and female zygotes XO in constitution. All females are basically $X^m O$ and the paternal X plays no part in sex determination. By selective non-disjunction the germ-line of the

Table 11. *Cases of Complementary Gametic Elimination.* (See text.)

	Species	Soma	Gametes ♀	Gametes ♂
Animals	1. *Enchytraeus lacteus* (2n = 170)	162 S + 8 L	162 S + 4 L	4 L
Animals	2. *Sciara coprophilla* (2n = 6 A + 2 Xp + 1 Xm)	♂ 6 A + Xm ♀ 6 A + Xp + Xm	3 A + X_m	3 A + 2 X_p
Plants	1. *Leucopogon juniperinum* (2n = 3x = 12)	AAB (12)	AB (8)	A (4)
Plants	2. *Oenothera*	aα	a	α
Plants	3. *Rosa canina* (2n = 5x = 35)	AABCD (35)	ABCD (28)	A (7)

female becomes $X^m X^m$ so that all eggs transmit an X^m chromosome. Functionally, therefore, the female remains the homogametic sex. Male zygotes are initially $X^m Y$ but the X^m chromosome is eliminated from the germ line which becomes YO in character (Ohno, Jainchill, and Stenius 1963; Ohno, Stenius, and Christian 1964). More recently Hayman and Martin (1966) have reported a different type of gonadosomic mosaicism in marsupials. The female situation is complicated by mosaicism within the soma; the skin, like the germ line, becoming XX as in *Microtus*. The male germ line retains the XY character of the male zygote but the Y is lost from part of the soma (Table 12). Significantly, as we shall (see pg. 54), a heteropycnotic Y may occur as commonly in the soma of the male mammal as a heteropycnotic X occurs in the female soma.

Complementary gametic chromosome elimination is known also in a few plants (Table 11). Here it has been exploited to allow numerically uneven polyploids to breed true via the sexual cycle. For example *Leucopogon juniperinum* is a permanent triploid of the type AAB and at meiosis 4 bivalents (AII) and 4 univalents (BI) regularly form both on the male and the female side ($2n = 3x = 12$). Functional pollen is haploid (A) and the univalents (B) are excluded from paternal transmission. Functional embryo sacs are diploid and carry the univalent elements (AB). Likewise the pentaploid *Rosa canina* is AABCD in constitution ($2n = 5x = 35$). The only functional pollen grains are haploid (A) while the embryo sac is tetraploid (ABCD).

The processes we have dealt with so far involve differences in chromosome number. SIGRID BEERMANN (1959, 1966) has, however, described a most remarkable instance of diminution in species of the copepod genus *Cyclops* which leads to differences in the heterochromatin content of germ-line and soma. For example, in *Cyclops strenuus divulsus* the diplotene chromosomes of the oocytes all possess long terminal segments of heterochromatin.

Table 12. *Gonadosomic Mosaicism in Mammals.*

	Species	♀		♂		Reference
		Soma	Germ Line	Soma	Germ Line	
Placentalia	1. *Microtus oregoni*	17 (XO)	18 (XX)	18 (XY)	17 (YO)	OHNO *et al.* 1963, 1966
Placentalia	2. *Acomys selousi*	XO	?	XY	?	MATTHEY 1965
Marsupialia	3. *Isoodon macrourus* 4. *Isoodon obesulus* 5. *Perameles nasuta*	leucocytes, liver and spleen 13 (XO), skin 14 (XX)	14 (XX)	leucocytes, liver and spleen 13 (XO), skin 14 (XY)	14 (XY)	HAYMAN and MARTIN 1966
Marsupialia	6. *Thalacomys lagotis*	18	?	19	?	

In developing embryos these segments are not distinguishable until telophase of the 4th cleavage. During the 5th cleavage division this heterochromatic material is expelled into the spindle. As a result the chromosomes undergo a drastic reduction in length and some of them show arm ratios which are different from any of those observed prior to diminution. This change is irreversible so that no heterochromatin is found in somatic cells after the 5th cleavage. The chromosomes in the primordial germ cells and its descendants, however, remain unchanged.

In *Cyclops furcifer* the chromosomes have heterochromatic segments at the medianly-located kinetochore regions as well as at the chromosome ends and this centric heterochromatin also undergoes diminution. Since the chromosomes remain V-shaped and are not broken by the diminution process it is clear that interstitial material can be removed from the chromosome without loss of its linear integrity. Finally this diminution process has been shown to involve an actual loss of DNA. Indeed, in terms of the percentage of total DNA present prior to diminution there is a 65% difference in DNA-content between post and pre-diminution stages. The male germ line, on the other hand, has the same DNA content as the female germ line.

b) Irregular Mosaicism and Chimaerism

A mosaic, as we have seen, is an individual with cell populations of more than one karyotype derived from a single zygotic karyotype through mitotic non-disjunction or mitotic elimination. In this it is distinguished from a chimaera which is an individual with cell populations of more than one karyotype arising through a mixture of different zygotic karyotypes following double fertilisation, chorionic vascular anastomosis or transplantation.

Table 13. *Sex-chromosome Mosaicism in Man.*

Pattern of Mosaicism	2n	Reference
1. XO/XX	45/46	Ford (1960)
2. XO/XXX	45/47	Jacobs *et al.* (1960)
3. XO/XX/XXX	45/46/47	Ferguson Smith *et al.* (1960)
4. XX/XXX	46/47	Ferguson Smith *et al.* (1960)
5. XX/XXY	46/47	Ford *et al.* (1959)
6. XY/XXY	46/47	Buckton *et al.* (1961)
7. XY/XXXY	46/48	Barr *et al.* (1962)
8. XXXY/XXXXY	48/49	Buckton *et al.* (1961)
9. XO/XY	45/46	Hirschhorn *et al.* (1960)
10. XX/XY	46/46	Gartler *et al.* (1962)
11. XO/XX/XY	45/46/46	Schuster and Motulsky (1962)
12. XX/XXY/XXYYY	46/47/49	Fraccaro *et al.* (1962)
13. XO/XYY	45/47	Jacobs *et al.* (1961)
14. XO/XXXY	45/48	Warkany *et al.* (1962)
15. XY/XXY/XXYY	46/47/48	McClean *et al.* (1962)
16. XXXY/XXXXY/XXXXYY	48/49/50	Gilbert-Greyfus (1963)

An extraordinary range of X and Y chromosome aneuploids have recently been discovered in man (see pg. 67). This, as we shall see, stems from the fact that although viable zygotes require at least one X-chromosome the epigenetic expression of any additional X-chromosomes appears to be quite small. A large number of mosaic types have also been reported in trisomic Klinefelter (XXY) males and in monosomic Turner (XO) and trisomic triplo-X (XXX) females as well as in the basic XY and the XX types (Table 13). A few mosaic types of the same pattern have also been found in the mouse. These mosaics result from post-fertilisation events involving either the simple loss of chromosomes or else their non-disjunction (Table 14). An equivalent mosaicism for sex-chromosome combination has, of course, long been known in *Drosophila* under the name of gynandromorphism. Here individuals possessing sharply defined cellular regions of different chromosomal constitution originate by the elimination of an X-chromosome from XX-individuals giving XX (♀) and XO (♂) sectors. Gynanders in *Bombyx mori,* on the other hand, are derived from binucleate eggs so that they are chimaeras. The normal female in *Bombyx* is XY in character so that eggs are normally X or Y and the complementary polar nuclei Y or X. In rare instances a polar nucleus may remain in the egg

Table 14. *Expected Sex-chromosome Anomalies Produced by First Cleavage Errors.*

Lost chromosomes or chromatids are shown in heavy type. The wavy line separates the 2 karyotypes present in mosaics. Similar events at later cleavages would produce mosaics of the same kind but ones in which the proportion of affected cells would be different; further, chromatid errors at one division are comparable with chromosome errors at the subsequent division.

(After RUSSELL 1961.)

Cleavage Error	Chromosome Loss	Chromatid Loss	Non-disjunction
X^mY Zygote	$\mathbf{X^m}$ Y → OY Lethal	$\mathbf{X^m}$ Y / X^mY → OY Lethal ∼∼ X^mY → X^mY ♂ normal	[X^m Y / X^m Y] → OY Lethal ∼∼ X^mX^mY → X^mX^mY Kleinefelter ♂
	X^m **Y** → X^mO Turner ♀	X^m **Y** / X^mY → X^mO ∼∼ X^mY ♀♂ Hermaphrodite	[X^m Y / X^m Y] → X^mO ∼∼ X^mYY ♀♂ Hermaphrodite
X^mX^p Zygote	$\mathbf{X^m}$ X^p → OX^p Turner ♀	$\mathbf{X^m}$ X^p / X^mX^p → OX^p ∼∼ X^mX^p ♀ Mosaic	[X^m X^p / X^m X^p] → OX^p ∼∼ $X^mX^mX^p$ ♀ Mosaic
	X^m $\mathbf{X^p}$ → X^mO Turner ♀	X^m $\mathbf{X^p}$ / X^mX^p → X^mO ∼∼ X^mX^p ♀ Mosaic	[X^m X^p / X^m X^p] → X^mO ∼∼ $X^mX^pX^p$ ♀ Mosaic

cytoplasm. Such binucleate eggs if fertilised by two sperm will come to possess two diploid nuclei, one male (XX) and one female (XY) in type. Sex chromosome chimaerism is known too in marmoset monkeys (BENIRSCHKE and BROWNHILL 1962, 1963) where it has been observed in both male (Table 15) and female cells in six out of fourteen animals representing five different species. The percentage of chimaeric cell admixture varied greatly while in non-hemopoietic cultures such as kidney and lung all the cells were normal. Marmosets usually have fraternal twins and there are regularly occurring anastomoses between the placentae of these twins. It would appear therefore that some primordial cells may gain access to the

Table 15. *Frequency of Chimaeric Cell Types in Male Marmoset Monkeys.* (After BENIRSCHKE and BROWNHILL 1962.)

Species	Karyotype		
	Marrow	Testis	
		Mitosis	Meiosis
	XY/XX	XY/XX	XY/XX
1. *Callithrix jacchus* (common marmoset)	49/11	22/5	3/0
2. *Cebuella pygmaea* (pygmy marmoset)	12/9	4/0	8/1
3. *Tamarinus mystax*	38/14	6/1	34/5

embryonic circulation and so be transported through vascular channels to the twin partner and become incorporated into both hemopoietic tissue and gonads.

A most remarkable example of irregular autosomal mosaicism has recently been described in the rainbow trout (*Salmo irideus*). Here, apparently, there is no fixed diploid number and the complement has been described by OHNO *et al.* (1965) in the following terms:

$$2n = 52 + \frac{\text{Number of Telocentrics}}{2}$$

A Robertsonian type of mosaicism exists within each individual, a certain number of chromosome arms undergoing fusion or fission in early embryonic life. In different tissues particular diploid numbers tend to predominate (Table 16) but the total number of chromosome arms remains fixed at 104. A somewhat similar state has been described by MATTHEY (1963) in *Acomys* while in *Peromyscus* mosaic variation is observed in chromosome arm lengths (see pg. 76).

A variety of mosaics and chimaeras of different kinds is known also in flowering plants. We have seen how, as a consequence of endoduplication, most angiosperms are a mixture of 2x–16x cells and are consequently mosaic. Following wounding, the hyper-ploid cells can come into division

Table 16. *Distribution of Modal Chromosome Numbers in Germ and Somatic Cells of three Different Individuals in each of four Age Groups in the Rainbow Trout, Salmo irideus.* (Data of Ohno, Stenius, Faisst, and Zenzes 1965.)

Age Group			Cell Type (2n/No. of Telocentrics)								Total Cells
			58 (12)	59 (14)	60 (16)	61 (18)	62 (20)	63 (22)	64 (24)	65 (26)	
(1) 3-week Old Eyed Embryo	Individual No. 1			2	**15**	6	1		3		27
	Individual No. 2			3	4		**8**	1	2		18
	Individual No. 3			1	**21**	2	1	1			26
(2) 1-month Old Fish	Individual No. 1	Liver			2	3	**13**	3	4		25
		Spleen		**9**	1						10
		Kidney		**6**							6
	Individual No. 2	Liver				2	**8**		2		12
		Spleen		1	**6**						7
		Kidney	**3**					1			4
	Individual No. 3	Liver			1	**14**	1			1	17
		Spleen	2	**4**							6
		Kidney		**8**							8
(3) 8-month Old Fish	Individual No. 1	Spleen		4	**6**	2					12
		Kidney	**8**	4							12
		Ovary	3	1	2	**5**	1	2			14
	Individual No. 2	Spleen		**7**	1	1				1	10
		Kidney	2	**4**							6
		Ovary			4		**8**				12
	Individual No. 3	Spleen		2	**8**						10
		Kidney		2	**4**						6
		Testis			**19**						19
(4) 18-month Old Fish	Individual No. 1	Spleen		**6**							6
		Testis			**4**		3				7
	Individual No. 2	Spleen		**8**							8
		Testis			3		2		3		8
	Individual No. 3	Spleen		**5**							5
		Testis		4	**9**						13

and may eventually form buds and shoots which are principally polyploid and which can be propagated vegetatively. In fact some of the first artificial autopolyploids were produced in this way (WINKLER 1916). The same effect can be obtained by colchicine treatment of the apex itself for this, initially, creates a mosaic condition between cells which are expected to divide anyway.

On occasion, however, mosaic buds do not "sort-out" and a euploid-hyperploid mixture persists in the apex. Now the apices of many plants are differentiated into a number of layers which are more or less distinct and independent of each other in that one does not contribute to another by periclinal cleavage. Shoots have thus arisen in which two or more of the histogens differ in chromosome number. And while some of these, radially-symmetrical, so-called periclinal, chimaeras break-down, many of them are very stable except when injured. In fact, the most elegant proof of the existence of self-propagating layers in shoot apices comes from work on polyploid chimaeras, principally those of *Datura* (SATINA 1959).

Some cultivated plants once thought to be polyploid have since proved to be cyto-chimaeras (EINSET and LAMB 1951). Thus in some apple varieties the three layers from the skin inwards are respectively $2x \rightarrow 4x \rightarrow 2x$ while others are $2x \rightarrow 4x \rightarrow 4x$. Both these combinations are stable while the sequence $2x \rightarrow 2x \rightarrow 4x$ is not. Again DERMEN (1953) finds that $2x \rightarrow 4x \rightarrow 4x$ and $4x \rightarrow 2x \rightarrow 2x$ colchicine induced periclinal chimaeras in *Prunus persica* are usually stable as are even more complex ones ($2x \rightarrow 8x \rightarrow 4x$ and $4x \rightarrow 2x \rightarrow 8x$) in *Datura*.

On the basis of the distinction made earlier, these cell mixtures should, strictly-speaking, be called mosaics. But the term chimaera is well established and the phenomenon is clearly different from the euploid-heteroploid mosaicism which arises during normal development. Further vegetative propagation allows for their indefinite perpetuation and indistinguishable conditions may arise in different ways. What is more, while grafting preceded the origin of some chimaeras, only the wounding necessitated by this operation appears to be important for their induction. Indeed the exact origin of many plant chimaeras is not known.

c) B-Chromosomes

In all plants and animals the karyotype includes a number of standard or A-chromosomes, every member of which has usually to be present to secure normal viability. The almost invariable properties of these A-chromosomes led the early cytologists to the generalisation that the chromosome number was necessarily constant within an individual and even a species. These A-chromosomes certainly represent a very delicately balanced system. Thus not only monosomics and trisomics (see pg. 68) but even deletion or duplication heterozygotes are frequently inviable or, at least, sub-vital.

In many species, however, the standard complement may be supplemented by varying numbers of supernumerary or B-chromosomes which, though evidently derived from the A-chromosome set, have become functionally subordinate to them. Indeed B's no longer pair with A-chromosomes and this is the principal property which distinguishes supernumerary members as B-chromosomes.

Where B-chromosomes occur they are found only in some individuals and are frequently smaller than the standard members of the complement (Fig. 23). There are, however, exceptions to this and in *Myrmeleotettix maculatus*, for example, the B-chromosomes are appreciably larger than

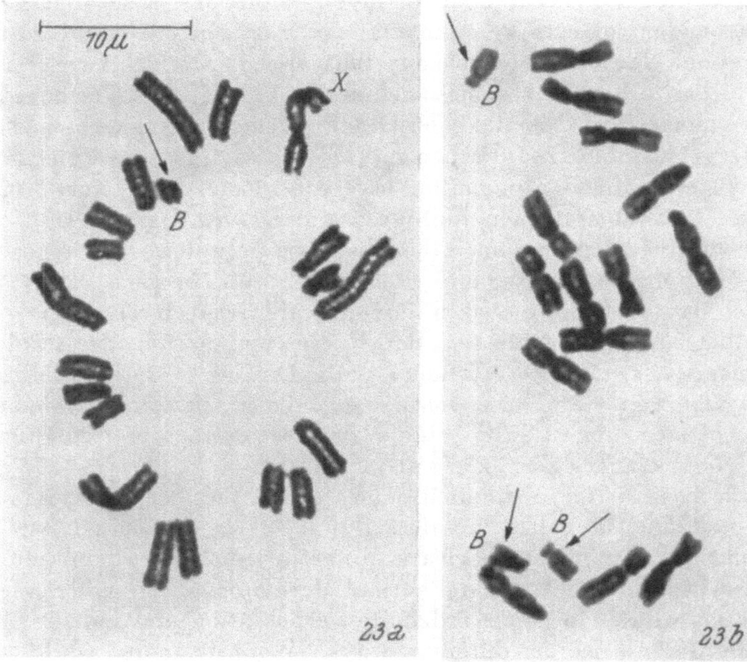

Fig. 23. Supernumerary (B) chromosomes at mitosis in *Pyrgomorpha kraussi*, 2n = 19 + 1B (Fig. 23a) and *Secale cereale*, 2n = 14 + 3B (Fig. 23b, kindly supplied by Dr. G. H. Jones).

many of the standard members (Fig. 24). The number of B-chromosomes per individual is usually low (1—3), especially in animals. In some plants, however, high numbers of supernumeraries have been found (Table 17) and even higher numbers have been introduced by artificial breeding.

The supernumeraries of many species differ from A-chromosomes in their behaviour also. Thus many supernumerary chromosomes appear to be mechanically less stable than the members of the standard set (Table 18). They are, in fact, subject to various kinds of maldisjunction. It is convenient and, perhaps, necessary to distinguish two kinds of situation in which their propensity for mechanical mis-behaviour may be expressed.

First, mitotic lagging leading to loss, or movement to one pole without separation of sister chromatids, may occur at unpredictable stages of development. The net result is the production of two new, numerically different, classes of nuclei in initially equal numbers and such mis-behaviour can be regarded as an uncontrolled anomaly. It may well be that B-chromosomes are no more susceptible than A-chromosomes in this respect, their apparent instability simply reflecting a lower pressure of selection against the new

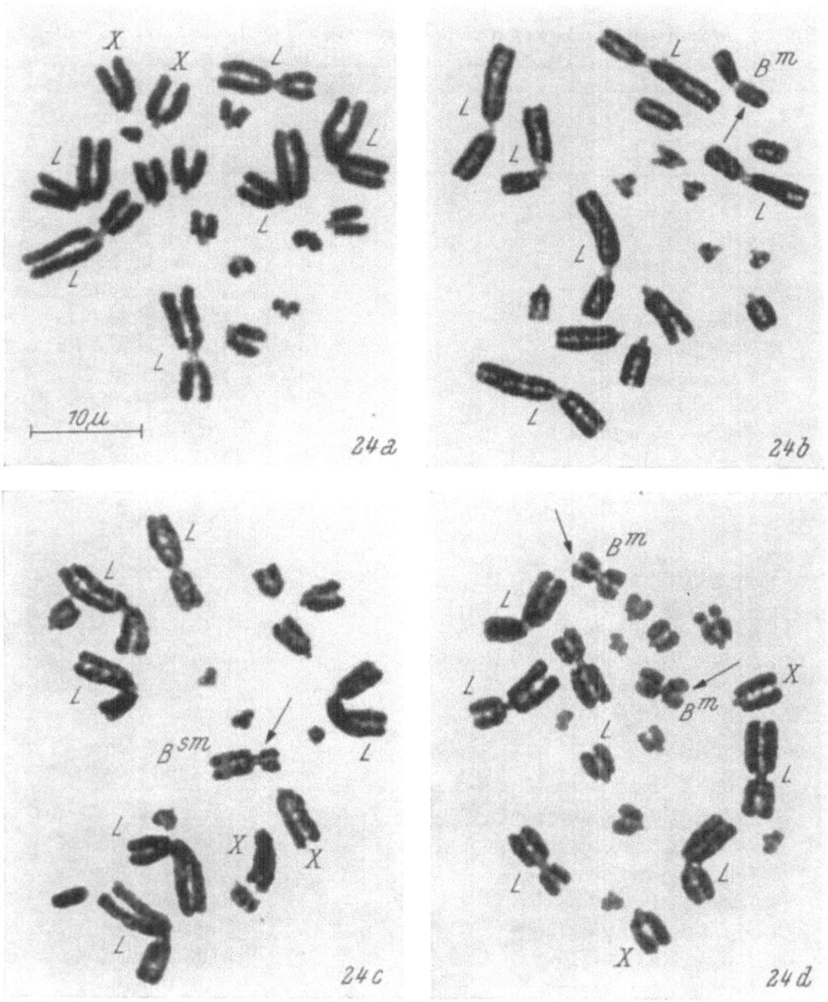

Fig. 24. Supernumerary (B) chromosomes from ovariole wall mitoses of female *Myrmeleotettix maculatus*. Fig. 24*a*, standard female complement, ♀ $2n = 16 + XX$: Fig. 24*b*, $18 + 1B^{m}$: Fig. 24*c*, $18 + 1B^{sm}$: Fig. 24*d*, $18 + 2B^{m}$ (see John and Hewitt 1965).

complements produced. Certainly stabilising selection in relation to A-chromosome anomalies is expected to be stronger. A rather exceptional type of instability has, however, been found in *Crepis capillaris* (Rutishauser 1960) where dicentric B's arise secondarily and produce bridges at meiosis.

On the other hand, the non-equational behaviour of B-chromosomes at mitosis in many plants is especially pronounced at particular stages of development so that its occurrence and frequency can be predicted with considerable accuracy. And, in this respect, they clearly differ from the standard chromosomes. For example, in *Sorghum purpureo-sericeum, Poa alpina, Poa timoleontis, Xanthisma texanum,* and *Haplopappus gracilis*

Table 17. *Maximum Numbers of B-chromosomes Recorded in Plant and Animal Species.*

	Species	Normal Diploid Count	Max. No. of B-chrms.	Reference
(A) Plants	1. *Centaurea scabiosa*	20	22	Frost 1957
	2. *Clarkia elegans*	18	6	Lewis, H. 1951
	3. *Crepis capillaris*	6	4	Rutishauser and Rothlisberger 1966
	4. *Crocus hyemalis*	6	4	Mather 1932
	5. *Dipcadi serotinum*	8	16	Resende and Franca 1946
	6. *Festuca pratensis*	14	9	Bosemark 1954
	7. *Lilium medeoloides*	24	11	Samejima 1958
	8. *Ornithogalum flavissimum*	12	5	Pienaar 1963
	9. *Paris tetraphylla*	10	9	Haga 1961
	10. *Rhinanthus major*	14	8	Wulff 1939
	11. *Rumex acetosa*	14/15	10	Haga 1961
	12. *Scilla autumnalis*	14	8	Battaglia 1964
	13. *Secale cereale*	14	10	Müntzing 1954
	14. *Solidago allisima*	54	5	Beaudry 1963
	15. *Tradescantia paludosa*	12	12	Anderson and Sax 1936
	16. *Trillium erectum*	10	9	Dyer 1964
	17. *Tulipa galatica*	24	12	Upcott and La Cour 1936
	18. *Viola rupestris*	20	8	Schmidt 1961
	19. *Viola montana*	40	10	Schmidt 1961
	20. *Zea mays*	20	34	Randolph 1914
(B) Animals	1. *Calliptamus palaestinensis*	23	4	Nur 1963
	2. *Diabrotica undecimpunctata*	19	5	Smith 1956
	3. *Helix pomatia*	54	6	Evans 1960
	4. *Locusta migratoria*	23	4	Itoh 1934
	5. *Myrmeleotettix maculatus*	17	3	John and Hewitt 1965
	6. *Nautococcus schraderae*	5	1	Hughes-Schrader 1942
	7. *Pseudococcus obscurus*	10	5	Nur 1962
	8. *Tipula paludosa*	8	4	Bauer 1931
	9. *Trimerotropis sparsa*	23	3	White 1951

there is a difference in the number of supernumarary elements found in the germ cells and certain of the somatic cells. Thus in *Sorghum* the B's are eliminated from all somatic tissues and retained only in the germ line. In *Poa alpina* they are omitted from adventitious roots as well as leaves and are present only in the central parts of the plant and in the germ cells. In many of these cases (Table 19) the variation in chromosome number is

Table 18. *The Incidence of Stable and Unstable B-chromosome Systems.*

B-type	Species	
	Plant	Animal
1. Stable	*Agrostis* *Alopecurus* *Anthoxanthum* *Briza* *Clarkia* *Dactylis* *Festuca* *Holcus* *Phleum* *Secale cereale*	*Circotettix undulatus* *Myrmeleotettix maculatus* *Pyrgomorpha kraussi* *Trimerotropis sparsa* *Diabrotica undecimpunctata* *Pseudococcus obscurus*
2. Unstable	*Crepis capillaris, conyzaefolia* and *pannonica* *Poa alpina, timoleontis* and *trivialis* *Haplopappus gracilis* and *spinulosum* *Sorghum purpuro-sericeum* *Xanthisma texanum*	*Calliptamus palaestinensis* *Camnula pellucida* *Helix pomatia* *Locusta migratoria* *Neopodismopus abdominalis* *Patanga japonica*

Table 19. *Distribution Patterns of Unstable B-chromosome Systems in Plants.*

Species	B-chromosomes			Reference
	Present		Absent	
	Constant	Variable		
1. *Crepis capillaris*	Rosette plant	Stem, bracts, younger inflorescences, receptacles and florets	—	RUTISHAUSER and RÖTHLISBERGER 1966
2. *Haplopappus gracilis*	—	Shoot	Root	ÖSTERGREN and FRÖST 1962
3. *Haplopappus spinulosum*	Root	PMC's	—	LI and JACKSON 1961
4. *Paris tetraphylla*	—	Root tips and ovules	—	HAGA 1961
5. *Poa alpina*	—	Central part of primary roots and in germ cells	Leaves and adventitious roots	MÜNTZING and NYGREN 1955
6. *Poa timoleontis*	PMC's	—	Roots	NYGREN 1957
7. *Poa trivialis*	Roots	PMC's	—	BOSEMARK 1957
8. *Sorghum purpureo-sericeum*	PMC's	Anther wall and ovaries	Roots and sterile flowers	DARLINGTON and THOMAS 1941
9. *Xanthisma texanum*	Shoots	—	Roots	BERGER and WITKUS 1954

Table 20. *Intra-individual Variation in the Supernumerary Chromosomes of Animals.*

Species and Reference	Individual No.	No. of Follicles with 0–4 B's					Total Follicles
		0	1	2	3	4	
1. *Calliptamus palaestinensis* (NUR 1963)	1	4		17	1		22
	2		31	12			43
	3		1	22	2		25
	4		20	12			32
	5		28	2			30
	6	3	2	30	2	1	38
	7		32	5			37
	8		6	25			31
	9	1	3	25			29
	10		4	37			41
	11		10	20			30
	12	3	5	25			33
	13	1	2	22	2		27
	14	4	12	23	1		40
	15	1	8	15			24
Totals	15	17	164	292	8	1	482
2. *Camnula pellucida* (CARROL 1920)	1	3	6	6			15
	2	2	2	8	1		13
	3	6		4	2		12
	4	1	11	3			15
	5		2	20			22
Totals	5	12	21	41	3		77

Species and Reference	Individual No.	No. of Cells with 0–6 B's							Total Cells
		0	1	2	3	4	5	6	
3. *Helix pomatia* (EVANS 1960)	1	4	33	3					40
	2		37	3					40
	3	4	52	8	1				65
	4			7	13	19	20	1	60
Totals	4	8	122	21	14	19	20	1	205

caused by somatic elimination so that the same number of B-chromosomes as originally transmitted by the zygote are maintained within the germ line.

Apart from mosaicism of this sort between germ line and soma, cases are also known where although both germ line and soma start off with the same number of B's the germ line does not transmit this number in its reduced form. Indeed the best studied cases of supernumerary misbehaviour are those involving polarised non-disjunction at mitosis in pollen grains and embryosacs. The polarity favours the movement of undivided B-chromosomes to the gametic nuclei or to those which give rise to gametes by equational division. Such behaviour leads to accumulation, inter-sib variation and to the preferential production of individuals with an even number of B-chromosomes. Variation and accumulation owing to non-disjunction at the premeiotic mitoses is also known in plants. Thus in *Crepis pannonica* Frost (1960) has found that individuals with 1 B in the root tip cells constantly have 2 B chromosomes in the pollen mother cells. An equivalent accumulation takes place on the female side in the turbellarian *Polycelis tenuis* through an endomitotic reduplication during the female meiotic prophase (Melander 1950). Likewise in *Crepis conyzaefolia* too there is a doubling of the supernumeraries in the germ cells and in Frost's (1962) opinion this doubling process is due also to a pre-meiotic endomitotic doubling of the B's. Animals have been investigated less thoroughly but in them also mitotic instability in the germ line has been recorded (Table 20) and this may provide a mechanism of accumulation (Nur 1963). Two other mechanisms are known to lead to the accumulation of B's. One operates at meiosis itself and depends upon preferential non-disjunction of univalent B's in the embryo sac mother cell (*Lilium, Trillium, Plantago*). The other involves directed non-disjunction of supernumeraries at first (*Secale, Collinsia*) or second (*Zea*) pollen mitosis and in the cases of *Secale* and *Collinsia* in the first mitosis of the embryo sac too (see Lewis and John 1963). Thus in various ways and for various reasons B-chromosomes, much more than members of the standard complement, are subject to considerable numerical variation both within and between sexual generations.

2. Epigenetic Mosaicism

In the above cases the cell variants within the individual differed genetically with regard to their chromosome complements. Where this condition develops regularly there can be no doubt that the changes are integral parts of the process of differentiation. For the most part, however, differentiation must be explained not on the basis of genotypic differentiation but on differential activity of undifferentiated genotypes. We have already encountered expressions of differential activity in the form of puff and loop development and regression (see pg. 6) and in the heteromorphism which is sometimes seen in the expression of secondary constrictions (see pg. 12). It is seen in an even more pronounced form however in the patterns of heterocyclic behaviour which characterise particular members of the chromosome complement.

a) Allosomes

The sex chromosomes frequently differ in behaviour from the autosomes with which they co-exist. Such differences may appear in the early stages of development when they subsequently characterise the somatic cells or they may be evident only at later stages and are then confined to the germ line. Alternatively both variants may co-exist in the same species. In all these cases the modified behaviour pattern takes the form of differential heteropycnosis.

i) Sex-Chromatin

In mammals the XX/XY sex chromosome system which characterises the majority of members may exist in a variety of forms (Fig. 25). The simplest and most frequent of these, and the one we may take as standard, is that where the single X of male somatic cells is invariably isopycnotic whereas in the female soma one of the two X-chromosomes is positively heteropycnotic, forming a prominent sex-chromatin body at interphase. Where, as in abnormal female karyotypes, three or more X chromosomes are present from the earliest cleavages the number of sex-chromatin bodies is always one less than the total number of X's. To accommodate this behaviour, as well as other observations, Lyon (1961, 1962) postulated that:

(a) The mammalian X when heteropycnotic is epigenetically inactive so that both males and females have, in effect, only one functional X-chromosome in somatic cells.

(b) This inactivation takes place at the time of implantation and is random in the sense that in some cells it is the maternal X which is inactivated while in others the paternal X becomes heteropycnotic.

(c) Heterochromatinisation occurs at only a restricted phase of development. The allocycly thus created, however, is persistent, even through successive mitotic cycles. Consequently, groups of cells derived from those originally subjected to the process of heterochromatinisation resemble each other and the parental cell with regard to their allocycly. Thus human male tumour cells of the type XXXY which have been secondarily derived from standard XY cells never show heteropycnotic X's because the single X of the male soma always behaves in an isopycnotic fashion and the three X's of the tumour cells are derivatives of this. Further, in cells which become endopolyploid the number of Barr bodies does not equal $x-1$ but $x(n-1)$ where x is the level of polyploidisation and n the number of X-chromosomes originally present. Thus the tetraploid cells of a normal female have two and not three Barr bodies while octoploid cells have four and not seven and so on. Evidently heterochromatinisation precedes polyploidisation.

Independent evidence for the physiological differentiation of the two-X-chromosomes in female mammals of the standard type is provided by the observation that the two X's have different patterns of DNA replication. In female somatic cells one of the two X's replicates its DNA late in the S-phase. Lyon's hypothesis thus implies that heterochromatinisation leading to the production of the sex chromatin body, late DNA-synthesis and genetic inactivity are all related properties.

Two tests have been made of this hypothesis of random inactivation:

(a) CATTANACH (1961) succeeded in inducing an X-autosome translocation stock of mice by treating wild-type CBA males with TEM (triethylenemelamine). The translocation was non-reciprocal involving the transposition of a segment of a medium sized autosome, represented by linkage

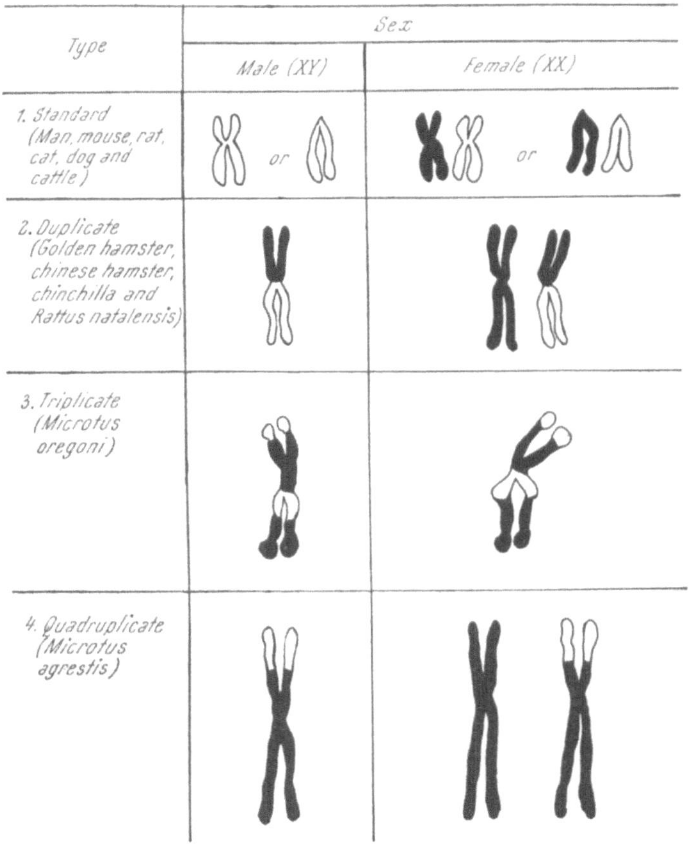

Fig. 25. X-chromosome form and behaviour in placental mammals. Those regions which are heteropycnotic in somatic nuclei are shown solid. In the case of the standard system the X is usually metacentric but may, as in the mouse, be telocentric (based largely on OHNO *et al.* 1964).

group 1, to the X. This led to the production of an X^T chromosome in which the autosomal segment was inserted into the X. The translocated sector was about one-third of linkage group 1 and contained the dominant wild-type alleles of the recessive loci chincilla (c^{ch}) and pink-eyed dilution (p). The addition of this segment to the X led to a marked increase ($\simeq 18\%$) in the length of the X^T over the X^N and, in consequence, while the X^N is no longer than the third longest autosome the X^T became the longest member of the complement and so was easily distinguished.

By suitable crosses it could be shown that $X^T X^N$ females and $X^T X^N Y$ phenotypically abnormal males showed a variegated phenotype, of the category termed flecked, with patches of wild type (grey) coat and patches of white (pc^{ch}) coat. But neither phenotypically normal $X^T Y$ males nor abnormal $X^T O$ females showed variegation.

Although the theory of X-inactivation was originally formulated on the basis of genetic data on the mouse, no sex-chromatin body has been demonstrated in this species. The reason for this is that sexual dimorphism of somatic interphase nuclei is obscured by the occurrence of several coarse chromocentres. These represent aggregations of the small blocks of heterochromatic material which lie adjacent to the centromeres in this species. Nevertheless the condensed X can be readily demonstrated in female prophase figures. Somatic prophases from the skin of flecked females of

Table 21. *Relationship of Individual X-chromosome Area to Total Diploid Chromosome Area in Some Mammals.*
(Data of OHNO, BEÇAK, and BEÇAK 1964.)
See also Fig. 25.

Species	2n	Total Diploid Chromosome Area (μ^2)	X-chromosome Area (μ^2)	X-autosome Ratio (%)
1. *Bos taurus*	60	165.73	4.65	5.60
2. *Canis familiaris*	78	160.36	4.11	5.07
3. *Mesocricetus auratus*	44	158.31	8.33	10.44
4. *Microtus oregoni*	17/18	152.28	12.70	15.71

the type $X^T X^N$ showed that the heteropycnotic X in white patches was larger than that in the dark patch. That is wild type patches contain a condensed inactive X^N and an isopycnotic X^T while white patches were populated by cells with a condensed and inactive X^T and an isopycnotic X^N. Equivalent observations were also made in $X^T X^N Y$ males where wild type patches were shown to carry a heteropycnotic X^N while in white patches the larger X^T was heteropycnotic. By contrast, in skin from wild type $X^T Y$ males and $X^T O$ females there was no condensed chromosome (OHNO and CATTANACH 1962).

(b) Autoradiographs of female mice of the type $X^T X^N$ show beyond doubt that the late labelling of the two X's is at random between cell lines for EVANS *et al.* (1965) found 92 "hot" X^T's to 88 "hot" X^N's. Similarly the two X-chromosomes of the female mule, one of which has been derived from the horse and the other from the donkey and which are also cytologically distinguishable, label in a random fashion (MUKERJEE and SINHA 1963).

Notice that the Y-chromosome of male somatic cells also shows positive heteropycnosity during interphase and is late labelling (YUNIS 1965). Notice also that in the three other types of XX/XY systems the pattern of heteropycnosity and late labelling is such that only 5% of the X: autosome ratio in each complement represents active X-material (Fig. 25, Table 21).

ii) Heteropycnosity in the Germ Line

Whereas in the homogametic sex of mammals sex-chromosome heteropycnosity is most marked in the soma, in the heterogametic sex their

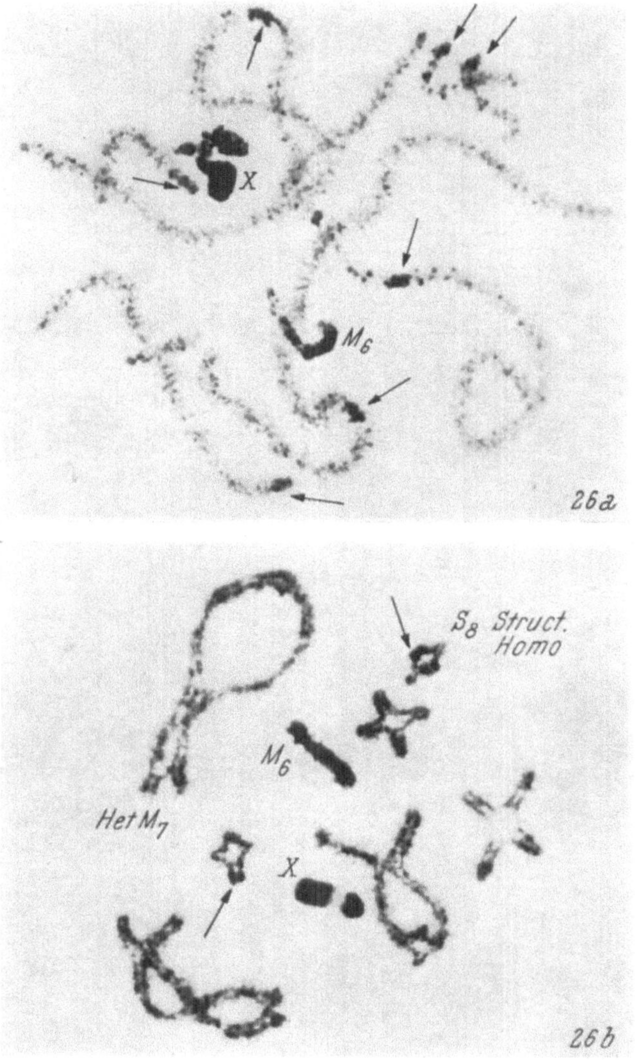

Fig. 26. Positive heteropycnosis at first meiotic prophase in truxaline grasshoppers. Fig. 26*a*, late zygotene in *Myrmeleotettix maculatus*. The univalent X and the M_6 bivalent are entirely heterochromatic and heterochromatic blocks exist around the kinetochores of the remaining autosomes (arrows). Fig. 26*b*, diplotene in *Chorthippus parallelus*. In addition to the X and the megameric M_6 pair which are regularly heteropycnotic the S_8 and M_7 bivalents are respectively homozygous and heterozygous for terminal supernumerary heteropycnotic segments (arrows).

pycnosity is more pronounced in the germ line. This heteropycnosis shows many variations with respect to the time at which it occurs and the extent to which it is carried. In many cases it first becomes manifest after the final spermatogonial division but other cases are known where it can occur in the early spermatogonia (see Fig. 3 *b*). Whether this is so or not, the sex

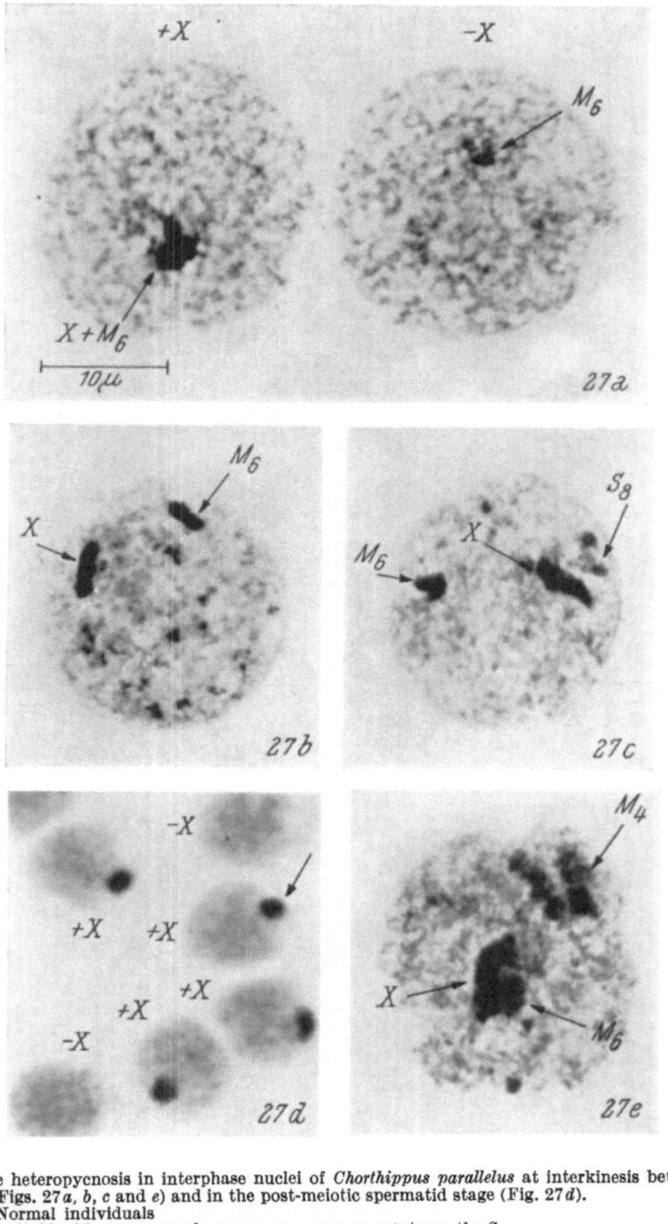

Fig. 27. Positive heteropycnosis in interphase nuclei of *Chorthippus parallelus* at interkinesis between first and second meiosis (Figs. 27 *a*, *b*, *c* and *e*) and in the post-meiotic spermatid stage (Fig. 27 *d*).
a, *b* and *d* — Normal individuals
c — Individual homozygous for supernumerary segments on the S_8
e — Individual tetrasomic for the M_4

chromosomes in the heterogametic sex frequently show positive heteropycnosis during the pre-meiotic period and then retain this character throughout the early stages of first prophase (Fig. 26). Sometimes they

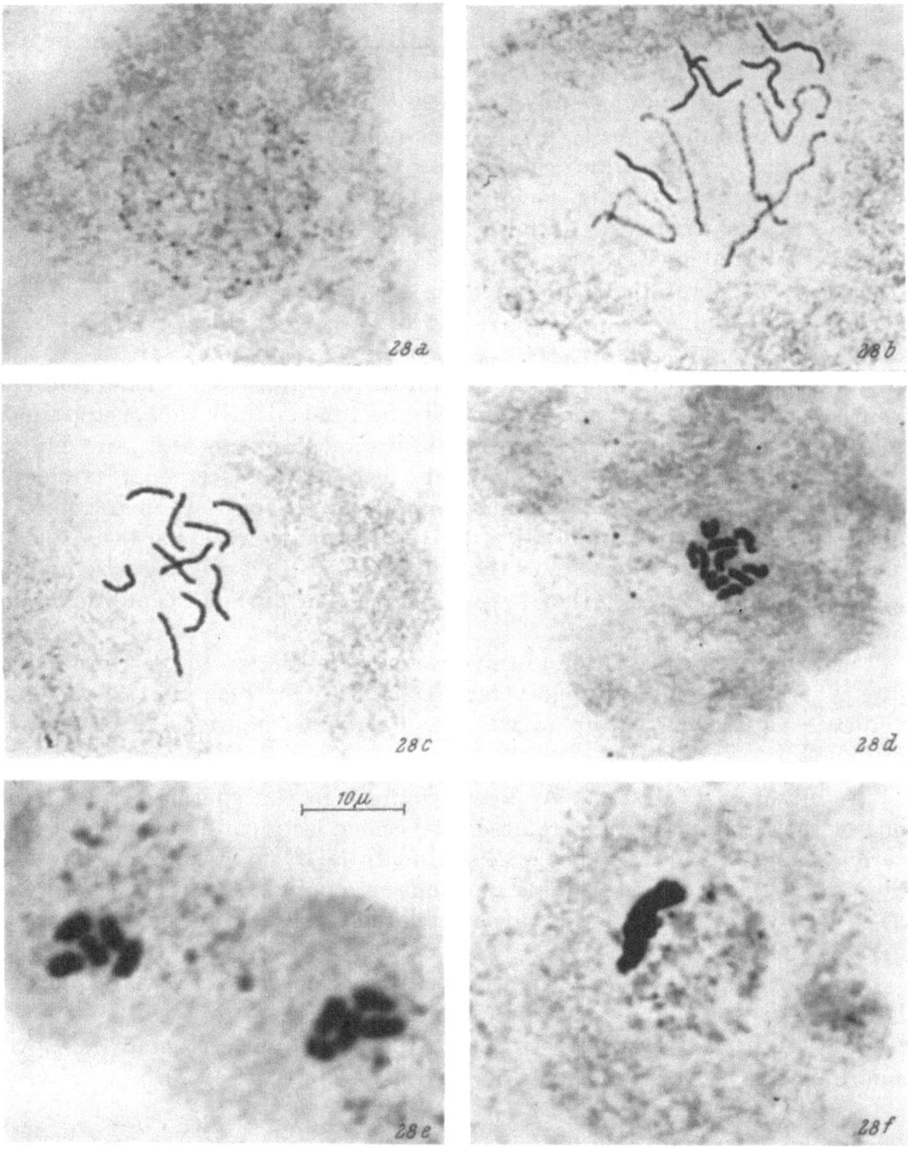

Fig. 28. Differential heterochromatic behaviour between homologous chromosome sets in the male of the mealy bug (2n = 10).
(*a*) Nucleus from a young embryo with minute scattered heteropycnotic regions. (*b*) Mid-prophase showing precocious condensation of the paternally derived set. (*c*) Late prophase and (*d*) metaphase of the same cell generation as (*b*) showing uniformity of condensation. (*e* and *f*) Nuclei of an older embryo in which the five allocyclic paternal chromosomes are respectively distinct (*e*) and aggregated (*f*). Photographs kindly supplied by M. Sabour and published with the permission of the editor of Science.

return to this state in the interphase between first and second division as well as the post-meiotic interphase (Fig. 27). Pycnosity at meiotic prophase in the heterogametic sex may, in some cases at least, be associated with the origin and maintenance of genetically differential segments.

b) Autosomes

The sex chromosomes are not the only members of the complement to show differential heteropycnosis. In many orthopterans there is a precocious "megameric" chromosome present in the complement which appears to develop its heteropycnosity at zygotene-pachytene (Fig. 25). Similarly the small m-chromosomes which characterise the coreid hemipterans show positive heteropycnosity at first meiotic prophase.

More pronounced cases are also known. Thus the E- and L-chromosomes already referred to (see pgs. 35 and 38) are also characterised by heteropycnosity. Perhaps the most extensive system on record is that found in certain coccid bugs (Brown and Nur 1964). In lecanoids, for example, all the paternally derived chromosomes become heteropycnotic at the blastula stage in male embryos and remain so throughout the whole of development (Fig. 28) and throughout meiosis too. The first meiotic division is equational, for the heteropycnotic and euchromatic chromosomes do not pair. During the second division the heterochromatic chromosomes segregate from the euchromatic and only nuclei containing the latter form spermatozoa. The net result is that only maternally derived chromosomes are transmitted by males which therefore behave as though they had been produced by haploid parthenogenesis although they are the products of diploid zygotes (compare *Sciara*).

A deviant sequence is found in *Comstockiella* (Brown 1963). Here too the paternal chromosomes become heteropycnotic at the blastula stage in male embryos and remain so up to meiosis. From this point on three variants are known:

(i) Standard-D system—At first prophase in the spermatocyte all but one of the heteropycnotic chromosomes become isopycnotic once more. One chromosome pair thus behaves in a differential fashion and this has been designated the D-pair. The single heteropycnotic chromosome (D^H) is a fixed entity and is eliminated at ana-telophase I while its homologue (D^E) divides equationally.

(ii) Variable-D System—Here different chromosome pairs may play the D-role but the D-pair is always cyst specific.

(iii) Multiple-D Variant—Here 1, 2 or 3 pairs may assume the D-role simultaneously. This variant is also cyst-specific (Nur 1965).

IV. Intra-Specific Polymorphism

Polyploidisation is a common occurrence in the somatic cells of animals and higher plants. With this exception the variations within the individual described in the previous section, while frequent or even regular in the types considered, are of sporadic occurrence when the whole animal kingdom is taken into account. This means that the germinal karyotype is normally transmitted faithfully so that the chromosome complements of two individuals can be compared without reference to variation within those individuals.

The range of variation revealed by comparisons between individuals is much wider than that observed within them. Thus the latter involves, for the most part, numerical variation and structural rearrangements are rarely encountered. Of course, differences which come to distinguish individuals may subsequently also serve to distinguish breeding groups but new chromosome types may also be retained within the original unit and thus contribute to intra-specific polymorphism. But whatever their subsequent fate these variants in the chromosome complement must originate within individuals.

1. The Origin of Mutation

Typically the chromosome complement in diploid organisms or diploid stages of organisms consists of two sets of identical partners. Structural differences can be found between homologous members as in the case of sex-

Table 22. *Spontaneous Chromosome Aberration Rates in Mammalian Cells in Vitro.* (See BENDER 1962.)

Species	Tissue	Cell Type	Breaks per Cell
1. *Ateles* (spider monkey)	Kidney	Epitheloid	0.014
2. *Cricetulus griseus*	Cell line	Fibroblast	0.040 – 0.210
3. *Homo sapiens*	Kidney	Epitheloid	0.009
	Blood	Leucocytes	0.022
	Skin	Fibroblasts	0.216
4. *Mus musculus*	Embryo	Mixed	0.01 – 0.35

chromosome heterozygotes (see pg. 93) or other regular states of heteromorphism (see pg. 73). Other structural differences appear from time to time as a result of chromosome mutation. The frequency with which these mutants arise is not generally well-defined nor do we have much information on the relative frequency of mutation in the germ line as opposed to the soma. One of the most useful places to study these problems is in mammalian cells grown in somatic tissue culture. Of course, such cells do not necessarily respond in the same way as cells *in vivo*. Indeed, as we shall see later (pg. 134), the abnormal environment in which cultured cells are maintained may lead to the production of abnormalities both in the structure and the number of chromosomes. With this caution in mind, rates ranging from 0.009–0.35 breaks per cell per generation have been reported for various cell types from various species (Table 22). Thus although the variation is usually low, values of 1.4% can occur.

Turning to the germ line we know from our own experience that in natural populations of grasshoppers entire germ-line mutants occur with a frequency of one in every two to three hundred individuals. Mosaic germ lines are, of course, much more common. In the australian grasshopper *Moraba scurra* WHITE (1964), from a study of some 17,000 individuals, has estimated that 1 in 750 individuals is likely to be heterozygous for a newly arisen rearrangement that is sufficiently obvious to be noted in a cursory examination. Likewise in *Purpura lapillus*, STAIGER (1959) reports that 11 out of 933 females examined were heterozygous for reciprocal translocations of different kinds, giving an overall frequency of approximately 1%.

Table 23. *Provisional Estimates of the Frequency of Newborn Individuals Showing Chromosome Aberrations in Man.*
(See also Table 31.)

Aberration Type	No. per 10^3 Live Born Children ♂	♀	Reference
1. Sex-chromosome anomalies	2.0	1.6	MACCLEAN, HARNDEN, COURT BROWN, BOND, and MANTLE 1964
2. Trisomy G (both simple and translocation trisomy)	1.5	1.5	PENROSE 1963
3. Trisomy (D + E)	0.7	0.7	MARSDEN, SMITH, and McDONALD 1964
4. Autosomal interchanges and pericentric inversions	5.0	5.0	COURT BROWN, JACOBS, BUCKTON, TOUGH, KUENSBERG, and KNOX 1966

Table 24. *Analysis of Chromosome Variation in Crepis.*
(Data of NAVASHIN 1926.)

Type of Individual			Species: *Crepis tectorum* (n = 4)			Species: *Crepis capillaris* (n = 3)	
1. Normal			3,957			1,989	
2. Numerical variants	Polyploid	3 x	16	39	43	11	12
		4 x	5			—	
		5 x	—			1	
	Aneuploid	2 x + 1	10			—	
		2 x + 2	4			—	
		2 x + 3	4			—	
3. Structural variants	Fragmentation		3	4		—	
	Translocation		1			—	
		Totals	4,000			2,001	

Human studies too (Table 23) and the little precise information available from plants (Table 24) give figures in general accord with the other estimates. They serve also to show that the incidence of numerical mutation is comparable with that of structural change.

2. Autosomal Polymorphism

a) Numerical

Typically chromosomes in diploid organisms or diploid stages of organisms (2 n) consist of two sets of identical partners (2 x). Reduplication of a whole set may lead to a nucleus with three, four or more sets, that is to a state of polyploidy (2 n = 3 x, 4 x ... etc.). Reduplication of some of the chromosomes of a set beyond the normal diploid number is called polysomy. The reciprocal results from the same series of errors as those which lead to polyploidy and polysomy are found in cells or in individuals which lack either whole chromosome sets (monoploids) or which lack individual chromosomes (monosomics). Polyploidy, in itself, involves no change in the numerical proportions of the chromosomes in a set; it is therefore a balanced change. Aneuploidy, however, whether it involves loss or gain of chromosomes, usually leads to an unbalanced state.

i) Polyploidy

Polyploidy within taxonomic species is very common in plants where it leads to the development of chromosome races (Table 25). In animals, on the other hand, diploidy is the rule. This is certainly not because the opportunity for the production of polyploids does not exist in animals. Polyploid cells, arising from failure of cytokinesis at premeiotic mitoses, are common, for example, in many insect testes (Fig. 29); indeed quite startling levels of polyploidy can be attained. And such states can give rise to hyperploid sperm (John and Henderson 1962). Where polyploidy does occur in animals it commonly arises in conjunction with parthenogenesis (Table 26). It is generally claimed that polyploidy occurs more readily in parthenogenetic forms because these secondarily lack those factors—normal meiosis and sex chromosome mechanisms—which are believed by many to constitute the major obstacles against the production of successful polyploids. Notice then that in *Tipula paludosa, Pales limulicornis* and *P. ferruginea* (Ullerich, Bauer, and Dietz 1964) triploid larvae which complete spermatogenesis have been found in laboratory stocks, and in the two species of *Pales* these have an XXY sex-chromosome constitution (Fig. 30).

Spontaneous and induced polyploidy has also been observed in isolated cases in the *Amphibia*. As a rule these polyploids are sterile. However, fertility has been reported in triploid and tetraploid axolotl females (Humphrey and Frankhauser 1946) and in both sexes of triploid *Triturus pyrrhogaster* (Kawamura 1951). In *Xenopus laevis* one 4 x and four 2 x/4 x mosaics were obtained from 427 tadpoles so that just under 0.3% were pure tetraploid. In urodeles the equivalent frequency is 0.04–0.4% according to the species (Fankhauser 1945).

Gurdon (1959) has reported that in *Xenopus laevis* tetraploid individuals can be fairly readily obtained by nuclear transplantation, the precise percentage varying from 5–13% according to the donor stage (Table 27). In external morphology the 4 x forms are almost identical with normal diploids. Their size is slightly smaller but their growth rate does not differ

Table 25. *Polyploid Races in Angiosperms.*

Species	Level of Ploidy								
	2x	3x	4x	5x	6x	7x	8x	10x	16x
1. *Agropyron duvalii*				35	42				
2. *Agropyron repens*			28		42				
3. *Agrostis stolonifera*	14	21	28	35	42	49	56	70	
4. *Allium nutans*	16	24	32	40	48	56	64		
5. *Alopecurus pratensis*			28		42				
6. *Anemone montana*	16		32		48				
7. *Bromus erectus*					42		56		
8. *Bromus inermis*					42		56	70	
9. *Bromus ramosus*	14		28		42				
10. *Calamagrostis brewerii*			28		42				
11. *Calamagrostis epigejos*			28		42		56		
12. *Callitriche stagnalis*	10		20						
13. *Crepis Bungei*	8		16						
14. *Festuca elatior*	14				42			70	
15. *Festuca ovina*	14	21	28		42	49	56	70	
16. *Festuca rubra*	14				42		56	70	
17. *Holcus mollis*	14		28	35	42				
18. *Phalaris arundinacea*			28	35	42		56		
19. *Potentilla opaca*	14		28						
20. *Prunus spinosa*	16	24	32	40	48				
21. *Ranunculus acris*	14		28						
22. *Salix aurita*	38		76						
23. *Silene ciliata*	24		48						192
24. *Solanum nigrum*	24		48		72				
25. *Themeda australis*	20	30	40	50	60				
26. *Urginea maritima*	20	30	40	50	60				
27. *Viola patrini*	24		48		72				

significantly from that of the diploids. The size of the organs is similar in diploid and tetraploid forms but the size of comparable 4x nuclei, and apparently of cells also, is twice that of 2 x forms. Tetraploids also have well developed secondary sexual characters and are indistinguishable from diploid types in this respect.

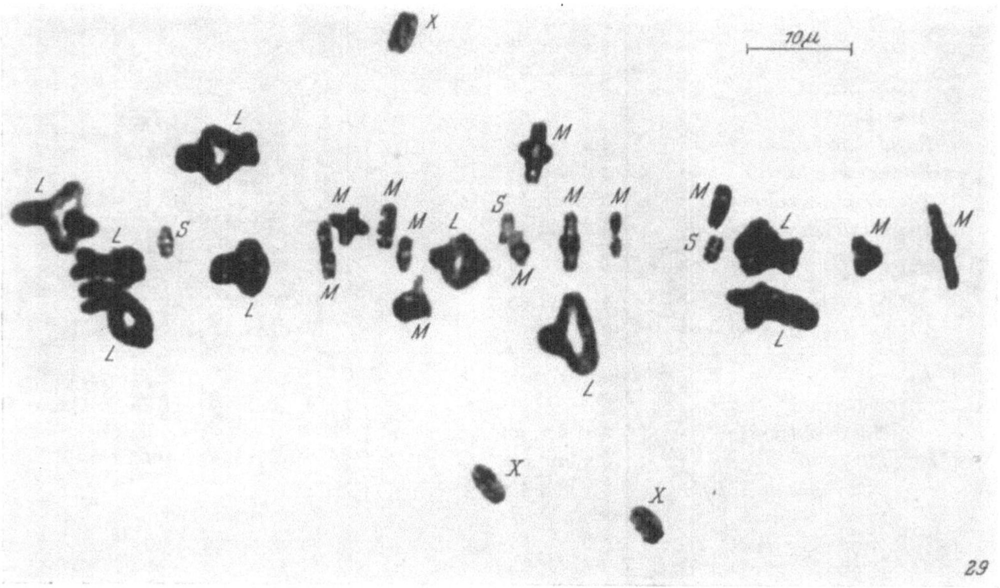

Fig. 29. A hexaploid cell (2n = 6x = 51 = 48 + 3 X) in *Myrmeleotettix maculatus* at first metaphase of meiosis. No multivalents are present because the state of hexaploidy has resulted from failure of pre-meiotic cytokinesis.

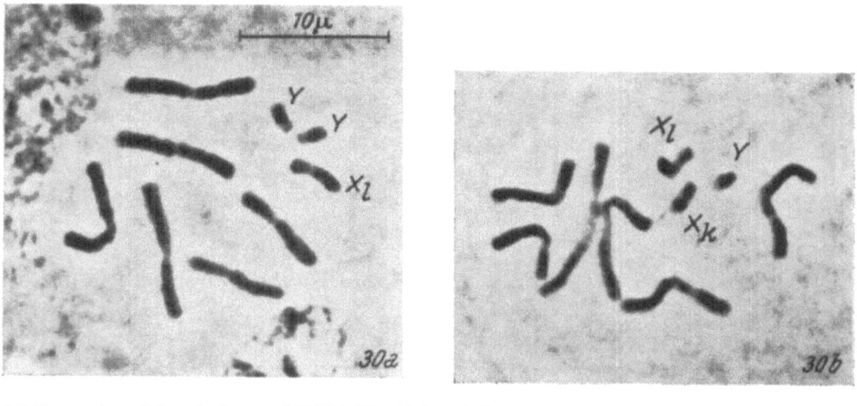

Fig. 30. Spermatogonial metaphases of XYY (Fig. 30*a*) and XXY (Fig. 30*b*) males of the crane fly *Pales ferruginea*. Notice that two morphological forms of the X exists, one short (X_k), the other long (X_l)—see also Fig. 52 (after ULLERICH, BAUER, and DIETZ 1964).

In tetraploid males the cells of the testis go through a perfectly regular first meiotic division but after this the nuclei are variable in size and shape and most of the spermatids are arrested as pycnotic nuclei. The remainder form spermatozoa of very variable size and shape and are incapable of effecting fertilisation presumably as a result of the irregular distribution

Table 26. *Polyploid/Parthenogenetic Female Races in Animal Species with Sexual Diploids.*

Group and Species	Type of Races	Reference
1. *Platyhelminthes*		
Dugesia benazzii	3 x soma/6 x germ line	BENAZZI 1966
lugubris	3 x soma/6 x germ line	
	3 x (soma and germ line)	
2. *Annelida*		
Buchholzia fallax	4 x	CHRISTENSEN 1961
Enchytraeus lacteus	4 x and 8 x	
Allolobophora caliginosa	3 x and 4 x	MULDAL 1952
Eiseniella tetraedratypica	3 x and 4 x	
3. *Crustacea*		
Artemia salina	4 x and 5 x	BARIGOZZI 1946
Trichoniscus elisabethae	3 x	VANDEL 1940
4. *Insecta*		
(i) *Coleoptera*		
Adoxa obscurus	3 x, 4 x, and 5 x	SUOMALAINEN 1958
Barynotus moereus	3 x and 5 x	
Otiorrhynchus scaber	3 x and 4 x	
Ptinus clavipes	3 x	SANDERSON 1960
Scepticus griseus	5 x	TAKENOUCHI 1961
(ii) *Diptera*		
Cnephia mutata	3 x	BASRUR and ROTHFELS 1959
Ochthiphila polystigma	3 x	STALKER 1956
(iii) *Lepidoptera*		
Solenobia lichenella and *triquetrella*	4 x	SEILER and SCHAEFFER 1941
5. *Amphibia*		
Ambystoma jeffersonianum	3 x	UZZELL 1963
6. *Reptilia*		
Chemidophorus exsanguis, rosselatus and *velox*	3 x	PENNOCK 1965

Table 27. *The Frequency of Tetraploids Obtained from Nuclear Transplantation in Xenopus laevis.*
(Data of GURDON 1959.)

Donor Stage	Number of Individuals		Percentage of Tetraploids
	2 x	4 x	
1. Late blastula	67	9	12
2. Gastrula	52	5	9
3. Neurula	22	1	4.5
4. Tail bud tadpole	16	1	6.25
5. Pre-hatching	47	4	8.5
6. Post-hatching	23	3	13

of chromosomes which is involved in their production. Likewise in 4x females the ovaries are reduced and immature and although the pre-meiotic cells look normal many oocytes become pycnotic. The *Xenopus* polyploids thus show that the main factor causing sterility is independent of the sex-determining mechanism and appears to be due to the general increase in chromosome number.

In mammals too triploid zygotes are not infrequent (Beatty 1957. Austin 1960) but unless these produce a sufficient number of diploid cells they die. Thus in rabbits (Melander 1950) and men (Böök and Santesson 1960; Ellis *et al.* 1963) the only viable triploids have been triploid/diploid cell mixtures. Male tortoise shell cats have likewise turned out to be triploid-diploid chimaeras with a 3x (XXY)/2x(XX) chromosome constitution (Chu, Thuline, and Norby 1964). In birds triploidy certainly appears to be compatible with healthy postnatal life, at least in the chicken. Thus the study of a fully grown Rhode Island Red with ambiguous sexual characteristics revealed a triploid constitution with 3A-ZZ in both soma and germ line (Ohno *et al.* 1963).

As far as monoploids are concerned there are many records of their sporadic occurrence both under natural and experimental conditions. Only one case is on record, however, of a vigorous monoploid. Daker (1966) has recently shown that the horticultural *Pelargonium* variety "Kleine Liebling" is a true monoploid. Aerial portions of the plant are invariably haploid while roots may be entirely haploid, haploid/diploid mosaics or purely diploid. The diploid cells appear to arise from the callus tissue which develops when new plants are initiated by stem cuttings.

ii) Aneuploidy

Autosomal aneuploidy in diploid sexual species is relatively rare, especially in animals, and it is customary to attribute this to the deleterious effects of the aneuploid state on development. This is in line with the principle of balance which is co-extensive with the qualitative differentiation of the A-chromosome set (see pg. 45). In the F_1 of a mutation experiment, in which wild type males were treated with TEM and then mated to untreated femals, Cattanach (1964) succeeded in producing a trisomic mouse ($2n = 2x + 1 = 41$). The mutant was detected only because it was completely sterile for it was otherwise phenotypically normal. The vitality of the trisomic must have been good for the animal was in good health when sacrificed at the age of 8 months. The testes were normal in size but few spermatogenic cells were found beyond the first meiotic division. Spermatids and spermatozoa likewise were all but absent and the few that were present were abnormal in shape. The sterility of the trisomic thus resulted from spermatogenesis breaking down after the first meiotic division. But the point of interest in the present connection is that this autosomal trisomy was not associated with any obvious morphological abnormalities. Meredith and Lyon (1966) have suggested that the extra chromosome was not a normal member of the complement but rather a translocation product. Thus if the material represented in triplicate was largely derived from

Table 28. *Sex Aneuploids in Man.*

Sex-chromosome Constitution	2n	Phenotype	Reference
XO	2x − 1 = 45	Turner syndrome	FORD *et al.* 1959
XXX	2x + 1 = 47	Female, variable phenotype	JACOBS *et al.* 1959
XXXX	2x + 2 = 48	Female, mental retardation	CARR *et al.* 1960
XXY	2x + 1 = 47	Klinefelter's syndrome	JACOBS and STRONG 1959
XXYY	2x + 2 = 48	Klinefelter's syndrome	MULDAL and OCKEY 1960
XXXY	2x + 2 = 48	Klinefelter's syndrome	FERGUSON SMITH *et al.* 1960
XXXYY	2x + 3 = 49	Klinefelter's syndrome	BRAY and JOSEPHINE 1963
XXXXY	2x + 3 = 49	Klinefelter's syndrome	FRACCARO *et al.* 1960
XYY	2x + 1 = 47	Male	SANDBERG *et al.* 1961

Table 29. *Zygotic Products of Normal and Non-disjunctional Segregation at Meiosis in Man.* Many of the types listed can also be produced by meiotic loss (XO) or cleavage errors (XO or XXX). Note the superscripts distinguish the origin of the two X-chromosome types.

Gametes	Male (X^mY)					
	Normal		Non-disjunctional 1st Division		Non-disjunctional 2nd Division	
Female (X^mX^p)	X^m	Y	X^mY	O	X^mX^m	YY
Normal X^p	X^mX^p normal ♀	X^pY normal ♂	X^mX^pY	X^pO	$X^pX^mX^m$	X^pYY
Normal X^m	X^mX^m normal ♀	X^mY normal ♂	X^mX^mY	X^mO	$X^mX^mX^m$	X^mYY
Non-disjunctional 1st Division X^mX^p	$X^mX^mX^p$	X^mX^pY	$X^mX^mX^pY$	X^mX^p normal ♀	$X^mX^mX^mX^p$	X^mX^pYY
Non-disjunctional 1st / 2nd Division O	X^mO	YO lethal	X^mY normal ♂	OO lethal	X^mX^m normal ♀	YY
Non-disjunctional 2nd Division X^pX^p	$X^mX^pX^p$	X^pX^pY	$X^mX^pX^pY$	X^pX^p normal ♀	$X^mX^mX^pX^p$	X^pX^pYY
Non-disjunctional 2nd Division X^mX^m	$X^mX^mX^m$	X^mX^mY	$X^mX^mX^mY$	X^mX^m normal ♀	$X^mX^mX^mX^m$	X^mX^mYY

the X then the absence of any obvious phenotypic effect would be perfectly understandable. However in their own irradiation experiments no X/A translocations were in fact recovered by MEREDITH and LYON although their investigation was undertaken principally for this purpose. Tertiary trisomics and interchange monosomics were nevertheless obtained.

Sex chromosome aneuploids are more common. Indeed an extraordinary range of X and Y chromosome types are now known in man (Table 28). The usual explanation offered for the production of these is meiotic nondisjunction (Table 29). As far as XXY types are concerned, studies using deutan colour blindness and Xg^a blood type as markers have shown that, in most cases, both X's come from the mother so that the individuals are $X^m X^m Y$ (FERGUSON-SMITH *et al.* 1964). It is commonly believed that extreme Klinefelter males (XXXY and XXXXY) are the products of double non-disjunction during oogenesis so leading to the production of female gametes with XXX or XXXX sex-chromosome complements. The available evidence, however, is that non-disjunction at the first meiotic division of the female is very rare. Likewise the low frequency of $X^m X^p Y$ Klinefelter males as compared with $X^m O$ Turner females suggests that in the male too non-disjunction at the first meiotic division is not common. RUSSELL (1961) has argued that since $X^m O$ is also much more frequent than $X^p O$ then there must be a relatively higher probability of loss of the X^p some time between fertilisation and completion of the first cleavage. OHNO, KAPLAN, and KINOSITA (1959) had in fact earlier concluded that non-disjunction must be more frequent at cleavage than at meiosis since they regularly found either a single X or a single Y in each of 1460 second meiotic nuclei of *Mus musculus*. And significantly $X^m O$ and $X^m X^p Y$ types have been demonstrated both genetically and cytologically in the mouse (RUSSELL 1961).

Supernumerary sex chromosomes are known also in several insects. Thus in the triploid fly *Pales ferruginea* ULLERICH, BAUER, and DIETZ (1965) have found a strain with XYY males

Table 30. *Supernumerary Sex Chromosome Types in Pales ferruginea.* (Data of ULLERICH, BAUER, and DIETZ 1965.)

Cross Type (♀ × ♂)	Progeny: Female		Male					Total
	2A + XX	2A + XXY	2A + XY	2A + XYY	2A + XXY	2A + XYYY	3A + XXY	
(2A + 2X) × (2A + XYY)	219	—	178	271	38	2	—	708
(2A + 2X) × (2A + XXY)	96	9	17	—	14	—	1	137

(Fig. 30). In crosses of this strain with normal XX females it was possible to recover XXY and XYYY males (Table 30). Both of these were fertile and here, as in man, there must be dominant or epistatic male determining factors in the Y. In natural populations of the gnat *Phryne cincta* too the diploid number shows considerable variation since 1—7 supernumerary Y chromosomes may be present (WOLF 1961). These have no apparent effect on morphology or sexual differentiation. Xyy and Xyyy males have been found also in laboratory strains of the coleopteran *Dermestes maculatus*

Table 31. *Classification of the Male Complement in Homo sapiens.*

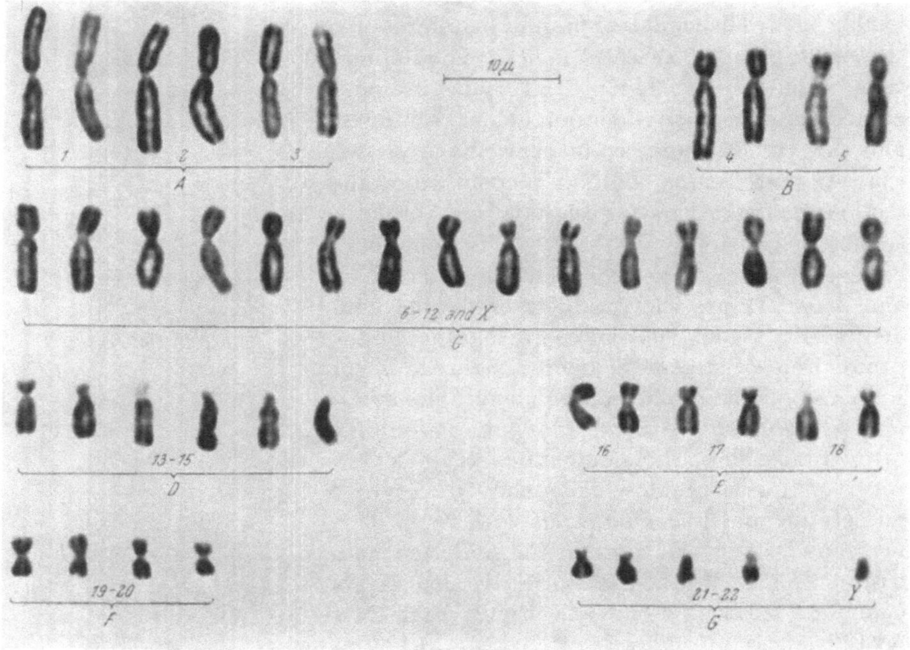

(see Fig. 50 b) while Xyy males are present too in *Dermestes frischii* (JOHN and SHAW 1967). Perhaps the most remarkable instance of supernumerary sex chromosomes is found in *Cimex lectularius* where up to 12 extra X-like chromosomes may be present in addition to the two X-chromosomes and the single Y normally present (DARLINGTON 1939). The extra chromosomes here are regarded as X-chromosomes because like them, and unlike the autosomes, they lack chiasmata in the male and divide equationally at first anaphase.

The wide range of X-chromosome aneuploids in man is undoubtedly due, in part at least, to the fact that the epigenetic expression of additional X-chromosomes is relatively small (see pg. 52). Likewise the Y-chromosome, which also replicates late, is involved in the production of sex aneuploids (XYY, XXYY and XXXYY) and in one case (XYY) the individual was phenotypically normal and fertile (SANDBERG *et al.* 1961). By contrast only 3 of the 22 autosome pairs in man appear to give viable trisomics (D — PÄTAU *et al.* 1960; E (18) — EDWARDS *et al.* 1960, and G — LEJEUNE *et al.* 1959 — see Table 31) and no viable monosomics have yet been found. Double trisomics, simultaneously triplo-18 and triplo-G are also known (GAGNON *et al.* 1961).

Table 32. *Chromosome Variation in Unbalanced Varieties of Hyacinthus orientalis.* Note diploid varieties have a complement including 8L ($2L^N$) + 4M + 4S = 16 chromosomes (see Fig. 22) while triploids have 12L ($3L^N$) + 6M + 6S = 24 chromosomes. (Data of DARLINGTON, HAIR, and HURCOMBE 1951.)

Type	Variety	2n	Complement		
			L (L^N)	M	S (S_2)
Hypo-triploids	Rosalie	17	9 (3)	4	4 (2)
	Cinderella	19	11 (3)	4	4 (2)
	Rosea maxima	20	10	5	5
	Van Speyk	21	10	6	5
	L'Orde Parfait	22	11	6	5
	Imperator	23	11	6	6
	President Roosevelt	23	12 (3)	5	6 (3)
	C. o. Haarlem	23	12 (3)	6	5 (2)
Hyper-triploids	Ostara	25	12 (3)	7	6 (3)
	Blue Herald	26	13	7	6
	La Peyrouse	27	14	6	6
	L'Innocence	27	15 (4)	6	6 (3)
	Schotel	27	15 (3)	6	6 (2)
	Q. o. Blues	27	15 (3)	6	6 (2)
	Mutant Q. o. Whites	27	15 (3)	6	6 (3)
	The Bride	27	15 (3)	6	6 (3)
	A. Arendsen	28	14	6	8
	Garrick	28	15	6	7
	Edelweiss	28	15 (4)	6	7 (3)
	Hoar Frost	28	16 (4)	6	6 (3)
Hypo-tetraploids	D. o. Westminster	29	15	7	7
	D. o. York	29	15 (3)	7	7 (3)
	Hindenberg	29	16 (4)	7	6 (3)
	Grace Darling	30	15 (3)	8	7 (3)
	Myosotis	30	15 (4)	8	7 (3)
	Totula	30	15	7	8
	Perle Brilliante	31	16 (4)	8	7 (3)
	Dr. Streseman	31	16 (4)	8	7 (4)

Significantly all these autosomal trisomics involve largely late replicating chromosomes. Indeed while it has been customary to regard the trisomic chromosome in Down's syndrome as number 21, the "mongolism" chromosome is late replicating and so is number 22 (YUNIS 1965).

The principle of differentiation which operates between the members of a chromosome complement, and which is co-extensive with the concept of balance, implies that different chromosomes have specific developmental qualities. In a few plant species, owing perhaps to a remote polyploid or hybrid ancestry, the chromosomes are not so differentiated. In such cases certain kinds of unbalance are hardly inferior to the normal state so that changes in number by gain or loss appear to have little effect. This appears to be the case in *Hyacinthus orientalis* under cultivation and in *Claytonia virginica, Cyrtanthus* (Ising 1962), *Erophila verna,* and *Narcissus bulbocodium* in nature.

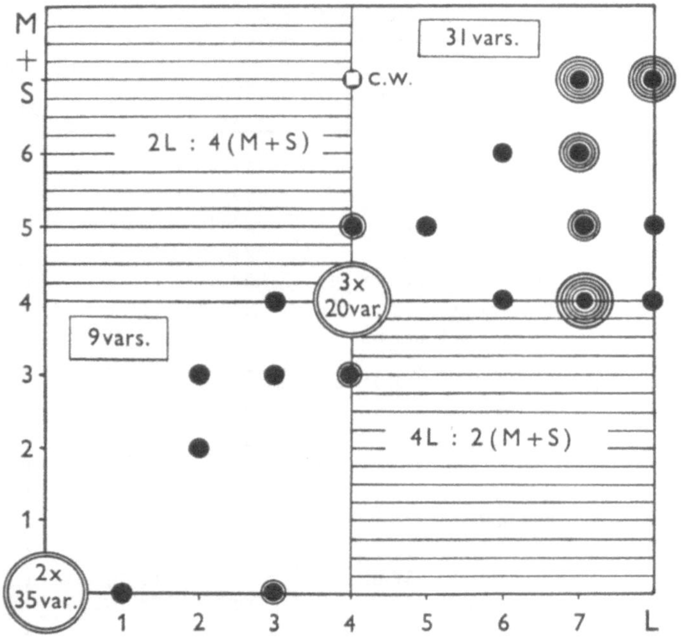

Fig. 31. The relationship between the number of extra long (L) and medium plus short (M + S) chromosomes in varieties of *Hyacinthus orientalis*. C. W. = var. Cardinal Wiseman (after Darlington, Hair, and Hurcombe 1951).

Hyacinthus orientalis is in its wild state a diploid species with 16 chromosomes comprising 4 long, 2 medium and 2 short pairs. The species was introduced into Holland in approximately 1560. The older varieties in cultivation have remained diploid (Fig. 22) but one of the results of cultivation has been the unconscious selection by breeders of triploid ($2n = 3x = 24$) and near tetraploid ($2n = 4x = 32$) forms (Darlington, Hair, and Hurcombe 1951). The triploids, unlike most triploids in flowering plants, are sexually fertile and in their progeny give a whole range of chromosome numbers from 17–23. These different numbers do not merely survive but, in addition, many of them have become successful garden varieties. Hyper-triploid and hypo-tetraploid varieties also occur (Table 32).

When the number of L-chromosomes relative to the number of M- and S-types are considered in the "unbalanced" varieties there is clear evidence of a selective concentration on certain combinations. Thus there is a preference for even numbers of M-chromosomes and odd numbers of L- and S-types. There is also a marked exclusion of 4L : 2 (M + S) and 2L : 4 (M + S) combinations (Fig. 31). The limits of "unbalance" are thus strictly set.

Narcissus bulbocodium has a similar disregard for balance for in addition to existing as a polyploid series (2 x → 6 x) in nature it also has a complete range of chromosome numbers from 2 x → 3 x (DARLINGTON 1963). *Erophila (Draba) verna*, likewise, has different races with a wide range of chromosome numbers (WINGE 1940). Diploids appear to have given rise to tetraploid and octoploid forms which, in turn, have secondarily produced a wide range of numerically unbalanced numbers. Indeed the original balanced forms of polyploid have themselves been supplanted while the diploid progenitor itself appears to have survived only in Denmark and Holland (Table 33). Variation of the same type and character has been described also in *Cardamine pratensis, Saxifraga granulata* and *Cochlearia anglica.*

Table 33. *Numerical Chromosome Variation in 113 Individuals of the Plant Erophila (Draba) verna.*
(Data of WINGE 1940 as summarised by DARLINGTON 1963.)

2n / Locality	2 x	4 x −	4 x +					8 x ±			
	14	24	30	32	34	36	40	52	54	58	64
1. Denmark	**33**	—	22	1	1	11	—	6	—	2	—
2. Holland	1	—	1	—	—	—	—	—	—	—	1
3. Britain	—	—	**11**	—	2	3	1	4	2	—	—
4. Germany	—	3	—	—	—	—	1	—	—	—	—
5. Sweden	—	—	1	—	—	**6**	—	—	—	—	—
Total	34	3	35	1	3	20	2	10	2	2	1

Table 34. *Numerical Chromosome Variation in 181 Plants of Claytonia virginica.*
(Data of W. H. LEWIS 1962.)
See also Fig. 32.

2n	No. of Plants	Totals
14	115	
15	1	123
16	5	
18	2	
25	1	
26	1	
27	1	
28	19	
29	19	57
30	5	
31	5	
32	3	
33	1	
36	2	
58	1	1

Table 35. *Chromosome Variation in 14 Aneusomatic Individuals of Claytonia virginica.* (Data of W. H. Lewis 1962.)

Plant No.	Chromosome Number																										Total Cells
	14	15	16	17	18	19	20	21	22	23	24	25	26	27	28	29	30	31	32	33	34	35	36	38	42	44	
1	3		1		4		9	2	4	3									2				1				29
2									5	1	2																8
3																1	2	6	2								11
4				1	3	2	2	1																			9
5	8	6	2	1																							17
6																2	6	4	1	1							14
7															8	3	1										12
8														2	8												10
9											6		9														15
10																		1	1					5	1	1	9
11																		1	3	2	3	1	1				11
12																			4	1			2				7
13									1		2	2	1	1	4	17	3	1	1								33
14														1		4	1		2		2	1					11
Totals	11	6	3	2	7	2	11	3	10	4	10	2	10	4	20	27	13	13	16	4	5	2	4	5	1	1	196

A particularly extreme instance of instability has been found in *Claytonia virginica* where fifteen distinct chromosome numbers were found among a sample of 181 plants (LEWIS W. H. 1962). Such diversity (Table 34) again involves the action of both polyploidy and aneuploidy. In addition, in a further 14 individuals the PMC's from the same bud differed in chromosome number, that is they showed aneusomaty (Table 35) superimposed upon the aneuploidy. LEWIS suggests that the unstable plants represent hybrids between individuals of different ploidy level. In *Hymenocallis calathinum* aneusomaty appears to be the rule for no constant chromosome number can be found in root tips. Numbers range from 23 to 83 (SNOAD 1955) but the lower, and consequently more unbalanced complements, are far less frequent than the higher ones. A similar though smaller range of variation (69 to 86) is found also in the PMC's. There would, therefore, appear to be different levels of tolerance to unbalanced chromosome numbers in anthers and roots, though the variation is certainly carried through the germ line.

Aneuploids are also frequent in apomicts, presumably because of the pronounced irregularities of chromosome behaviour that can occur in the EMC's. Thus in the genus *Poa* the bulk of the members seem to be aneuploid. For example the following ranges have been recorded (GUSTAFSSON 1947):

Poa pratensis	$2n = 48 - 124$
P. subcaerula	$2n = 82 - 147$
P. arctica	$2n = 39 - 92$
and *P. alpina*	$2n = 31 - 57$

These forms must be quite insensitive to chromosome unbalance. In *Agrostis stolonifera* and *Holcus mollis* chromosome counts made on seedlings raised from pentaploid and hexaploid races are also mainly aneuploid both after open-pollination and intercrossing. Yet in large samples from natural populations only a very small number of such types have been detected (JONES 1957).

b) Structural

Structural mutations may, if they involve chromosomes of different type or segments of different length, change the morphology of the chromosome complement. Even where they fail to do so they can usually be detected at meiosis because they lead to the development of novel patterns of pairing. They have the same effect on the polytene homologues of dipteran flies. Our present interest in these mutations, however, is not in their pairing behaviour but rather the changes they effect in the karyotype.

i) Inversion Differences

The order of the genetic material in a chromosome may become inverted and the inversion may (peri-centric) or may not (para-centric) involve the centromere. Paracentric inversions produce no change in either centromere position or overall length and they can therefore be detected only where the chromosome arm is adequately labelled cytologically as in polytene chromosomes or chromosomes with pronounced chromomere patterns. Easily detectable modifications of the complement generally occur after pericentric inversion because, unless the breaks are equidistant from the centromere, arm ratios are altered. Similar changes in relative arm lengths can, of course, be effected by translocations involving 3 breaks and reunions. But no practical demonstration of this has ever been given in natural situations.

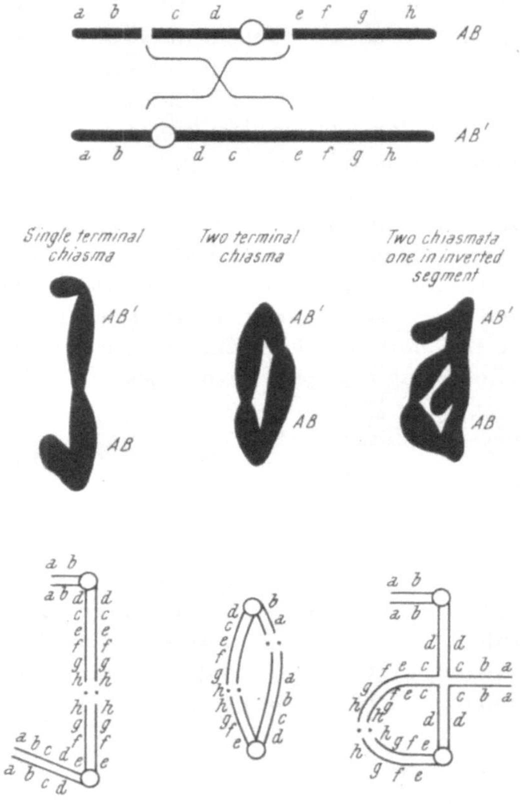

Fig. 32. The nature and behaviour of a spontaneous pericentric inversion heterozygote involving the AB metacentric of an individual (Michelago 955) of *Moraba scurra* (based on White 1961).

Some of the best known instances of pericentric inversion are found in grasshoppers. For example in the eumastacid grasshopper *Moraba scurra* the large metacentric AB chromosome normally has two limbs which are almost equal in length. White (1961) found one individual where one of the AB chromosomes was normal while the other had one arm twice the length of the other but showed the same overall length (Fig. 32). Regular polymorphisms for such pericentric inversions exist in natural populations of *Moraba scurra* (CD and EF chromosomes—White 1956), *Moraba virgo* (CD chromosome—White, Cheney, and Key 1963), *Moraba viatica* (A and CD chromosomes—White, Carson, and Cheney 1964) in Australia and in North American species belonging to the genera *Trimerotropis, Circotettix,* and *Aerochoreutes* (White 1959, 1951). Polymorphism for pericentric inversion is known also in the bark weevil genus *Pissodes* (Manna and Smith 1959). More recently a number of cases of pericentric inversion have been uncovered in rodents. Thus Matthey (1963, 1966) has demonstrated polymorphisms for pericentric inversions in *Mastomys natalensis* and *Mus minutoides minutoides* respectively. Likewise autosomal polymorphism for presumed pericentric inversions have been described for the largest chromosome pair in wild roof rats (*Rattus rattus*) from Japan (Yosida, Nakamura, and Fukaya 1965) and in the third chromosome of laboratory and wild specimens of *Rattus norvegicus* from Misima and Kusudomari (Yosida and Amano 1965). In both species the chromosomes concerned exist both as telocentrics and acrocentrics of the same overall length.

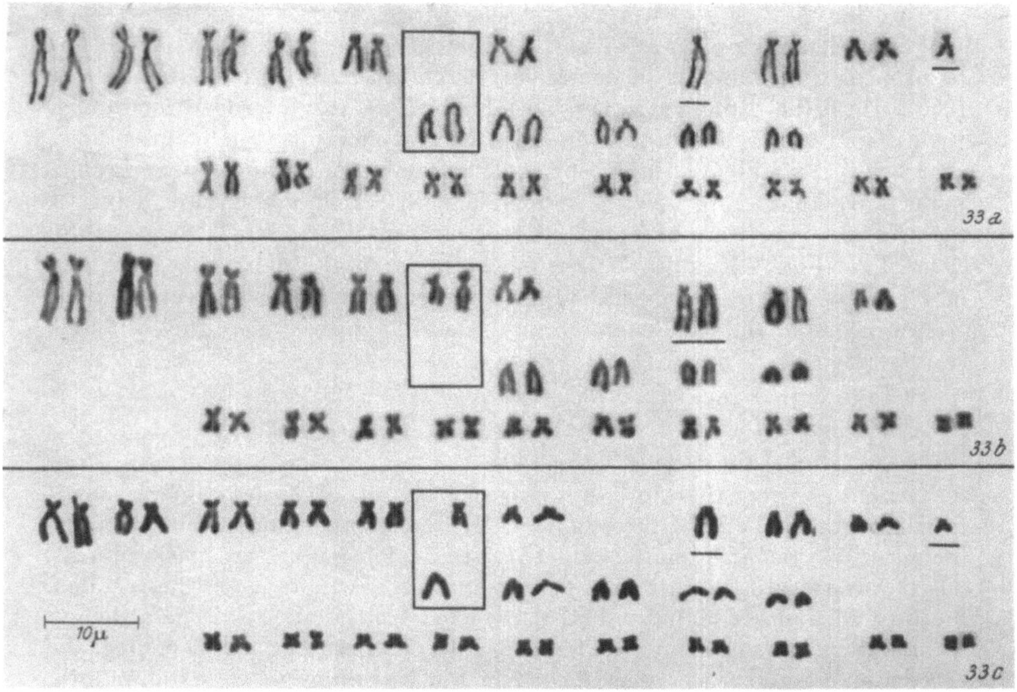

Fig. 33. The production of an F_1 male heterozygous for a pericentric inversion (Fig. 33*c*) following the mating of homozygous parents (♂, Fig. 33*a* and ♀. Fig. 33*b*) in *Peromyscus maniculatus bairdii*. The X and Y chromosomes are underlined while the autosomal pair involved in the polymorphism is enclosed by a rectangle (after OHNO *et al* 1966).

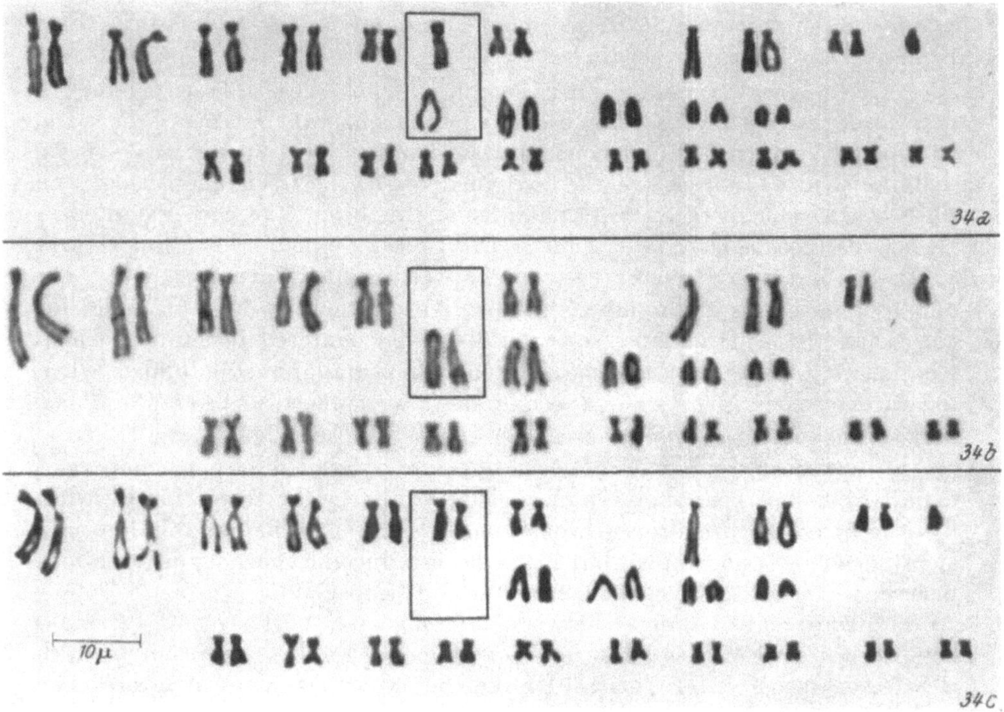

Fig. 34. The chromosome complements of three cell types obtained from the spleen of one F_1 male of the type illustrated in Fig. 33*c* (after OHNO *et al* 1966).

Perhaps the most interesting example of pericentric inversion so far reported is that found in the deermouse *Peromyscus maniculatus* (OHNO et. al. 1966). Here the diploid number is 48 but the number of rod chromosomes varies from one individual to another. At first metaphase of meiosis, however, every animal invariably forms 23 autosome pairs and a sex chromosome pair. In a cross between a female with 8 rods and a male with 10, nine F_1 offspring were obtained all heterozygous with respect to one autosome pair (Fig. 33) and, as expected, both parental types were recovered in the F_2 generation. Parental types were, however, also recovered in the spleen of one of the F_1's where an analysis of 23 mitoses gave the following results (Fig. 34):

19 cells with 9 rods (F_1 type),
3 cells with 10 rods (paternal type),
1 cell with 8 rods (maternal type).

It would appear therefore that new somatic cell types are being reconstituted within the body of the cytologically heterozygous individual although the germ line remains unaffected. This presumed somatic segregation parallels the findings in the rainbow trout (see pg. 43) where, it will be recalled, the diploid number of individual fish varied from tissue to tissue although the number of arms remained fixed at 104. In *Peromyscus,* of course, the chromosome number remains the same but variation is observed in arm lengths.

ii) Interchange Differences

Exchanges between chromosomes can likewise produce detectable alterations in the chromosome complement. Such inter-changes fall into three principal categories:

(a) Reciprocal exchanges between non-homologues. These produce a detectable change in the chromosome complement only in cases where the exchanged segments differ markedly in length. Two such cases can be distinguished. First, where the exchange occurs between one V and one rod-shaped element (Fig. 35). Examples of this type have been recorded in *Allium fistulosum* (ZEN 1961) and *Chorthippus brunneus* (JOHN and HEWITT 1963). Second, where the exchange is between the long arm of one element and the short arm of the other, irrespective of whether these are meta- or acrocentric types (Fig. 36). NODA (1960) has encountered just such a translocation in *Lilium maximowczii.* A somewhat similar result follows when the chromosomes involved in the exchange are markedly different in size, as was the case in *Disporum sessile* (KAYANO 1960, see Fig. 37).

(b) Reciprocal exchanges between opposite arms of different members of a pair of homologous metacentrics. This results in the formation of what have been called pseudo-iso-chromosomes (Fig. 38). Whether this happens spontaneously is not known but it has been induced experimentally both in barley (CALDECOT and SMITH 1952) and maize (KOO 1958).

(c) Reciprocal exchanges between the long and short arms of two non-homologous acrocentrics leads to a very unequal kind of interchange which is referred to as centric *fusion.* Unlike the two other types of interchange this may produce not only a structural but also a numerical alteration in the complement. Thus if the small centric portion is lost there will be a reduction in the chromosome number by one though the total number of major arms remains unaffected (Fig. 39). The term acrocentric was intro-

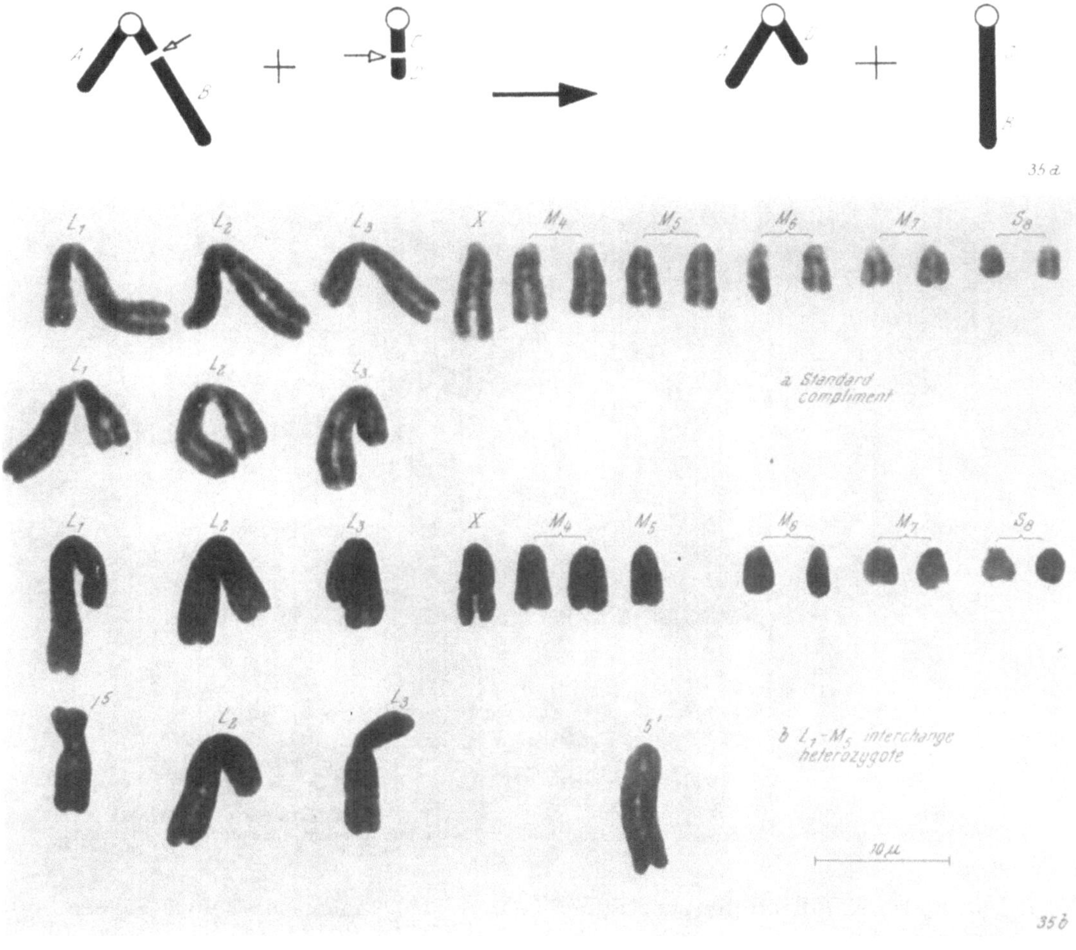

Fig. 35. The theoretical (Fig. 35*a*) and practical (Fig. 35*b*) consequences of an interchange between a meta- and a telo-centric chromosome. The karyotypes shown in Fig. 35*b* represent respectively the standard and the mutant form of *Chorthippus brunneus*.

duced by WHITE (1945) as an alternative to the historically older term telocentric because this author believed that all chromosomes were fundamentally two-armed. This has led to considerable confusion in the use of the terms acro- and telo-centric. We have already qualified our own intentions (see pg. 12) but it may be as well to re-state them here. Telocentric means having a strictly terminal centromere while acrocentric means having a subterminally located centromere. A number of the early cytologists, especially those of the American school, believed that permanent centric fusions could result from direct adhesion between the chromosome ends of rod-shaped elements. HELWIG (1941), for example, argued that the metacentrics present in short-horned grasshoppers were equipped with two centromeres situated very close together. There is certainly evidence that the kinetochores of telocentric entities may engage in non-random association (MARKS 1957). There is also evidence that irregular and temporary fusion of telocentrics may take place and that such fusions may persist

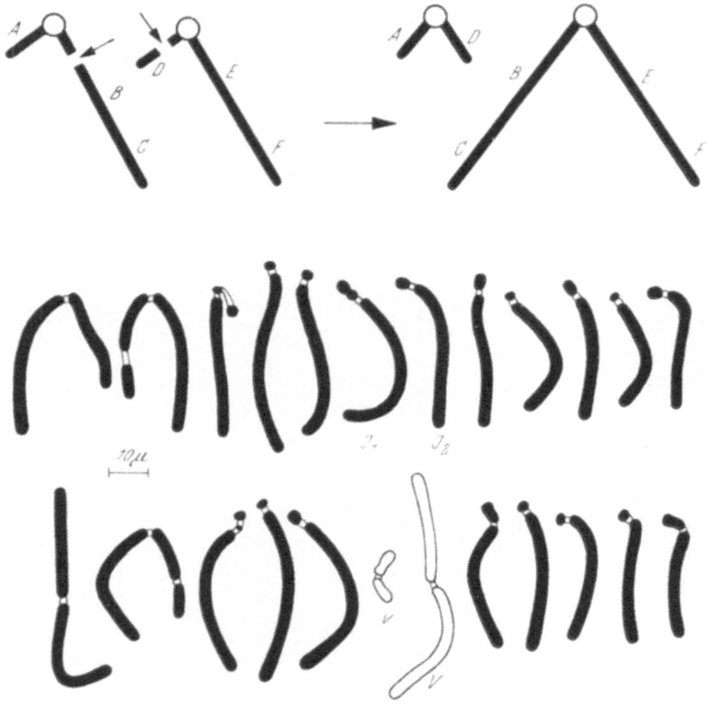

Fig. 36. The theoretical (upper) and practical (lower) consequences of an interchange between a short and a long arm of two acrocentric chromosomes. The karyotypes in the lower Fig. represents the normal and the mutant form of *Lilium maximowiczii* (after NODA 1960).

through several mitotic generations (WOLF 1960). Such cases, however, are probably not to be compared with regular and permanent centric fusion and most contemporary authorities reject the idea of direct fusion between unbroken ends. Even so there are two further possibilities which could also lead to an equivalent state. First, fusion following breakage in both of the short arms of two acrocentrics or else breakage and fusion within the centromeres themselves, a process which could apply equally well to both acro- and telo-centric systems.

It has, however, been the practice to regard fusion as a process involving the loss of one centromere and the concomittant conversion of two acrocentrics into one metacentric element. This practice undoubtedly stems from the widespread tendency to view centromeres as indivisible, telomeres as irreplaceable and telocentric chromosomes as rare and predominantly unstable. In recent years, however, an increasing number of chromosome complements have been described in animals in which many or indeed all the members are telocentric (see pg. 112) and the significant point is that

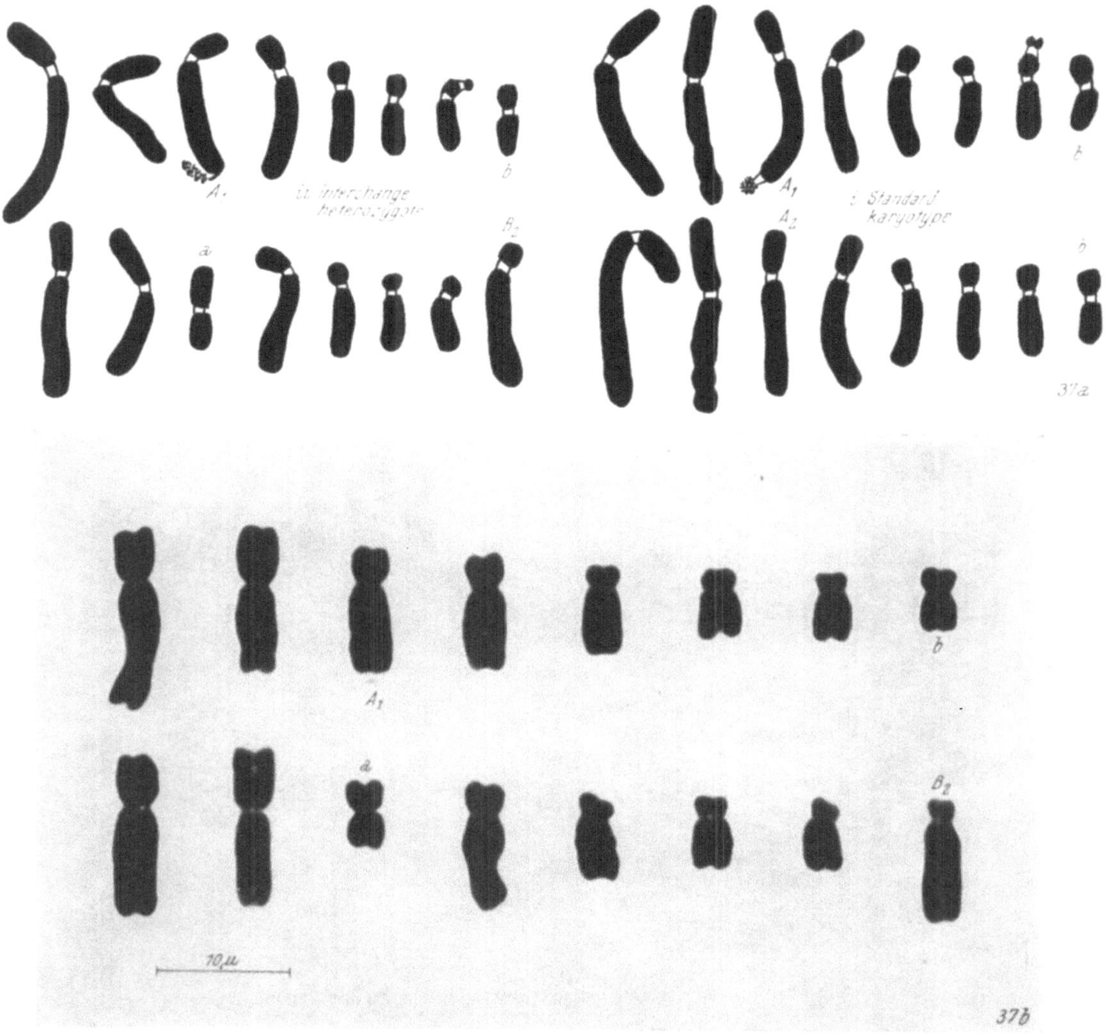

Fig. 37. A spontaneous reciprocal translocation between a large (A_2) and a small (*b*) chromosome in *Disporum sessile* (after KAYANO 1960; the preparation used to illustrate the heterozygous state in 37*b* has been made from material kindly supplied by Dr. HIROSHI KAYANO). Note, the standard complement includes two heteromorphic pairs.

in complements of this kind metacentric elements may arise both in culture and *in vivo*. For example GUSTAVSSON and ROCKBORN (1964) described a divergent complement in three cases of overt lymphatic leukaemia in cattle. Here, instead of the 60 telocentric elements normally present, only 59 chromosomes were found and one of these was a metacentric which represented a fusion between the largest and the smallest members of the complement. The same abnormality was also found in a foetus from one of the affected cows so that the condition is clearly transmissable through the germ line. More recently GUSTAVSSON (1966) has examined 1,134 animals from SLB and SRB swedish breeds of cattle. Of these 89% had the normal complement

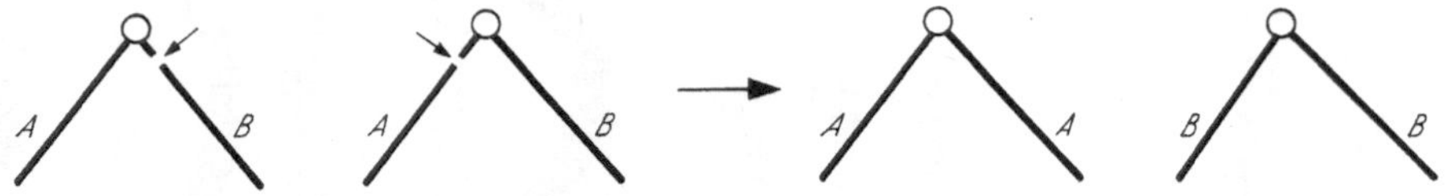

Fig. 38. The production of pseudo-isochromosomes following interchange between different arms of homologous chromosomes.

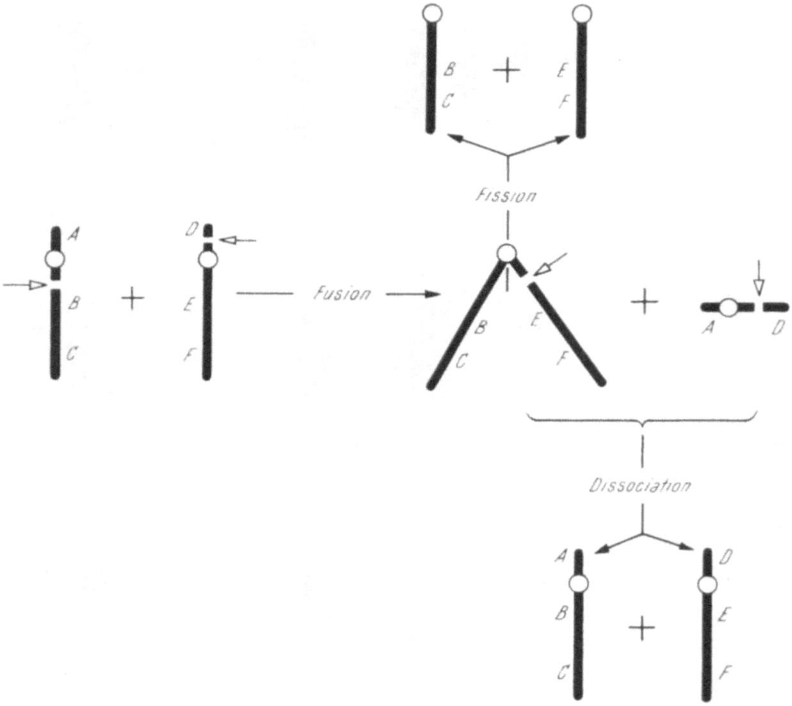

Fig. 39. The inter-relationship of fusion, fission and dissociation.

of 60 telocentrics, 129 animals had 59 chromosomes including one metacentric while 4 had 58 with two metacentrics. An equivalent situation has been described in Saanen goats (SOLLER, WYSOKI, and PADEH 1966).

MORRISON (1954) recovered an interchanged chromosome in the progeny of a monosomic wheat plant. Since the interchange involved the same chromosomes as the misdivision which gave rise to the telocentrics, MORRISON suggested that the interchange itself arose from the fusion of two telocentrics from non-homologous chromosomes (Fig. 40).

Of course, fission through the centromere (misdivision) or else dissociation as a result of a markedly unequal translocation between a large metacentric and a small donor chromosome (WHITE 1957b) can, in theory, lead to a similar Robertsonian relationship (Fig. 39). For this reasons many cytologists prefer simply to state that two rod-shaped chromosomes correspond to one V-shaped element without speculating on the direction of

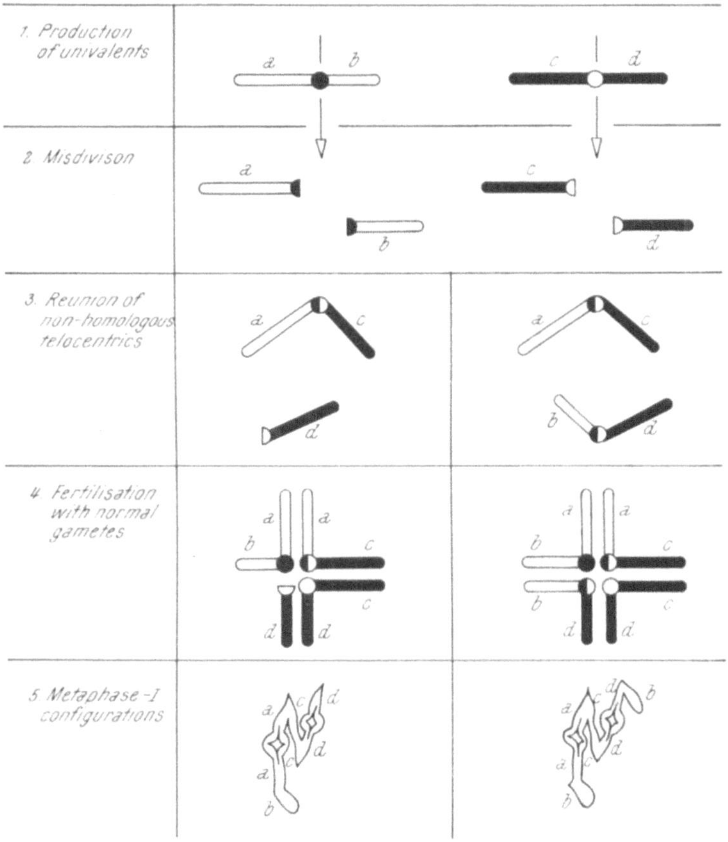

Fig. 40. The production of interchange types following the reunion of telocentrics formed by misdivision (based on MORRISON 1954).

evolutionary change or the process involved. There is of course a basis for distinction—fission gives rise to telocentrics while both fusion (*sensuo stricto*) and dissociation must involve acrocentric elements. The problem is thus essentially one of distinguishing between acro- and telo-centric chromosomes. WHITE has repeatedly claimed (1954, 1957) that in animals the centromere never occupies a strictly terminal position. This view is no longer tenable (JOHN and HEWITT 1967 and see Figs. 41 and 42); indeed telocentrics may well prove to be common in certain groups (see pg. 112). But to err on the side of caution, polymorphic relationships between rod and V-shaped chromosomes of the type implied above are well known in the *Orthoptera* (Figs. 43 and 44). They also occur in certain mammals (Table 36). In rodents, for example, chromosome numbers are very variable and apart from differences in number between different species, which we shall deal with later (see pg. 110), three cases of Robertsonian relationships are on record:

(i) Eurasian mice, for example, show a great uniformity of karyotype with 2 n = 40 rod-shaped elements. By contrast pygmy mice from different parts of Africa not only show marked differences between species but also chromosomal polymorphism within species. Thus in *Mus minutoides minutoides* 2 n = 18 or 19, in *M. minutoides musculoides* 2 n = 31, 32, 33 or 34 and in *M. triton* 2 n = 20, 21 or 22. In the latter case MATTHEY (1963) finds the range of chromosome numbers to be correlated with a polymorphic relationship between rod and V-shaped elements (see Table 36). At meiosis in the 21 chromosome type a "trivalent" association is formed indicative of a fusion/fission relationship.

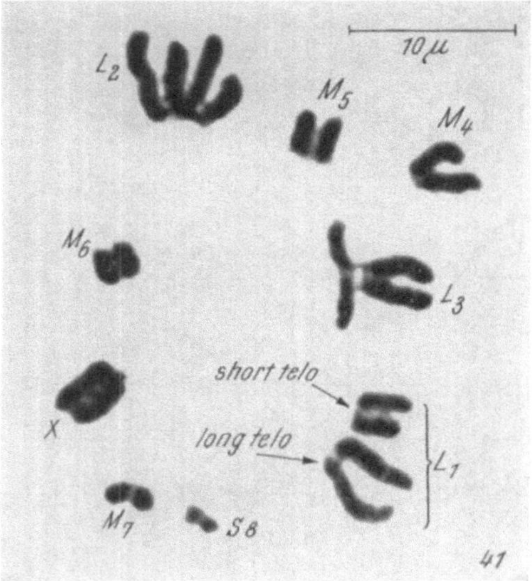

Fig. 41. The production of telocentrics at metaphase-II following the misdivision of the L_1 metacentric in the grasshopper *Myrmeleotettix maculatus.*

(ii) The common shrew, *Sorex araneus,* occurs as two distinct chromosome races. One of these, found in France and Belgium, is monomorphic and is characterised by a male complement of 23 chromosomes (XY_1Y_2) and a female complement of 22 (XX). The other race is of wider distribution occurring in Germany, Denmark, Finland, Norway, Sweden, Holland, France and Britain. Populations of this race show Robertsonian polymorphisms involving autosome pairs 3 to 8. This leads to a range in chromosome number from 21 to 33 in the male and from 20 to 32 in the female (MYELAN 1965). The variable elements may be represented by a single metacentric or else by two rods. As expected "trivalent" associations develop in primary spermatocytes of heterozygous individuals (FORD, HAMERTON, and SHARMAN 1957).

(iii) In *Gerbillus pyramidum* there exist three allopatric chromosome races which are not recognised taxonomically. One, in Algeria, has 40 chromosomes, a second in the coastal plains of Israel has 52 chromosomes and a third in Negev has 66. Their relationships (Table 36) suggests a Robertsonian scheme and this is supported by observations on a male hybrid produced by crossing a 66♀ with a 40♂ (WAHRMAN and ZAHAVI 1955). At

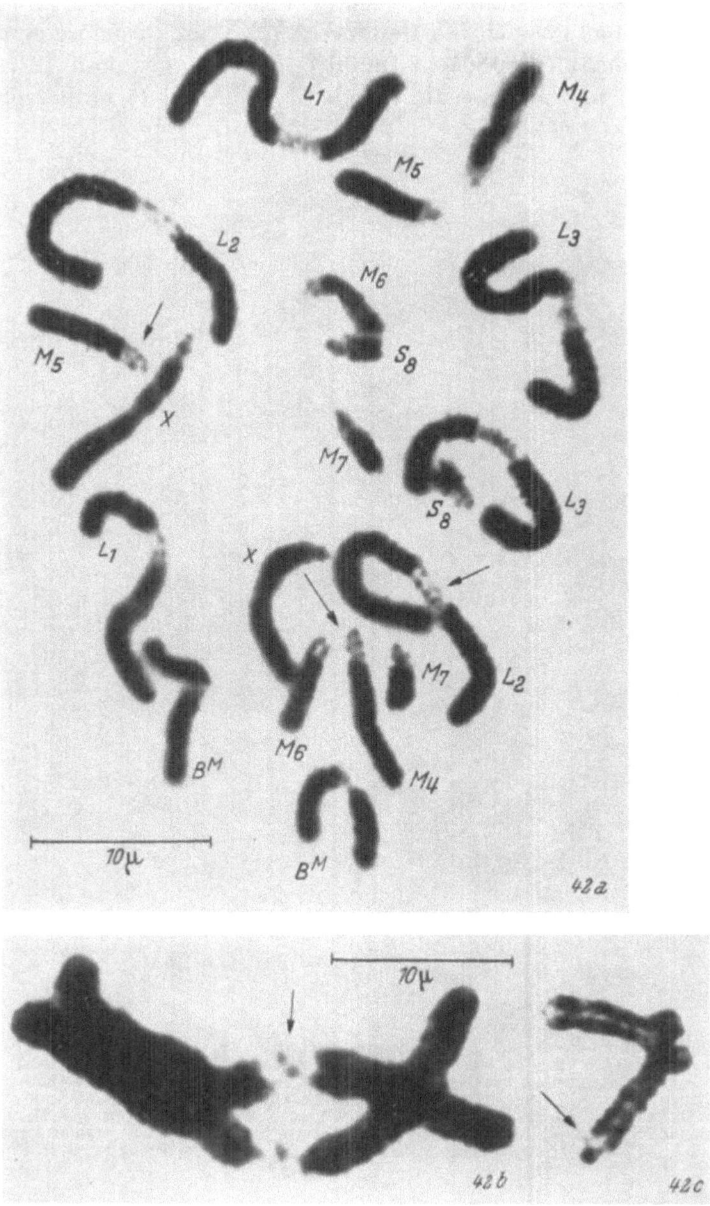

Fig. 42. Centromere organisation at mitosis (Fig. 42*a*, ovariole wall c-mitosis of *Myrmeleotettix maculatus* $2n = 16 + XX + 2B^{m}$) and meiosis (Figs. 42*b* and *c*, diakinesis of *Chorthippus brunneus*) in truxaline grasshoppers.

meiosis this gave 3 II + (8–10) III + (3–4) C (V–VII). The formation of trivalents bears witness to the homology of rod and V-shaped chromosomes in the two populations while the presence of compound chains of V–VII elements suggests that certain of the metacentrics in the two genomes are homologous for only one of their arms.

One case of special interest is that which exists in pigs. The domestic pig has 38 chromosomes with a standard XY/XX sex chromosome mechanism and the autosomal complement includes 24 V-shaped and 12 rod-shaped elements. Most of the domestic breeds are believed to be descended from

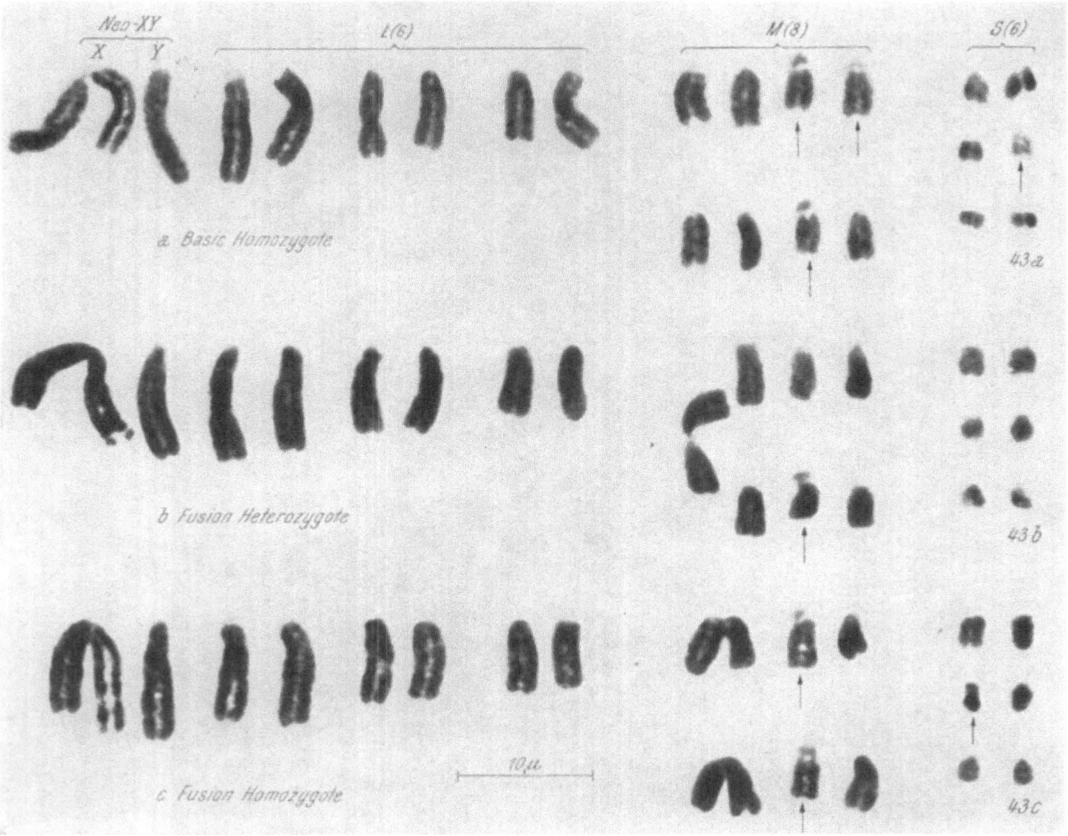

Fig. 43. Polymorphism for centric fusion in the male of the north american grasshopper *Oedaleonotus enigma*. The standard karyotype (*a*) has 2n = 20 + neoXY in which there are 6 long, 8 medium and 6 short autosomes. The fusion involves two of the medium members, probably the $M_4 + M_5$, and can be found in both heterozygous (*b*) and homozygous (*c*) states. Notice that heterozygosity and homozygosity for pericentric inversions (arrows) occur commonly in the M_6, M_7 and S_9 members. Thus (*a*) is homozygous for an inversion in M_6 and heterozygous for an inversion in M_7 and S_9. (*b*) is heterozygous for inversions in M_7 while (*c*) is heterozygous for inversions in M_6, M_7 and S_9. (These preparations were made from material kindly supplied by Dr. G. M. HEWITT.)

the european wild pig. Out of 36 wild pigs examined by MCFEE, BANNER, and RARY (1966) twenty six had a complement of 36 members including 26 V and 8 rod autosomes. The remainder possessed 37 chromosomes including 25 V and 10 rod autosomes. The authors concluded that pure strains of wild pig probably have 36 chromosomes and that the presence of 37-types results from breeding between wild and domesticated forms in their mixed herd. This case is of importance on two counts. First, it is one of the few instances where the direction of change is known. Since domestics pigs

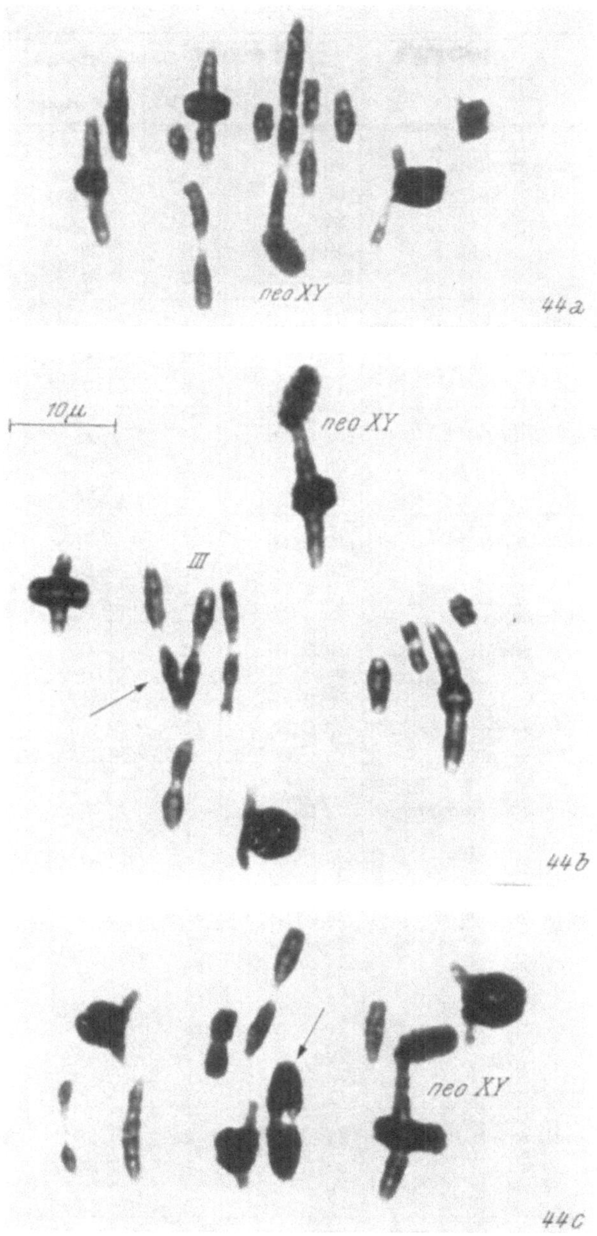

Fig. 44. First meiotic metaphase in the basic homozygote (*a*), the structural heterozygote (*b*) and the structural homozygote (*c*) of *Oedaleonotus enigma*. The fusion heterozygote and homozygote are indicated by arrows. (These preparations were made from material kindly supplied by Dr. G. M. Hewitt.)

are derivatives we must be dealing here with a case of centric fission. The second point of importance is that at least two of the 37-chromosome forms were fertile.

Table 36. *Polymorphic Relationships Between Rod and V-shaped Chromosomes within Species.*

Group	Species	2n	Autosomes		Reference
			V's	Rods	
Mollusca	*Purpura lapillus*	36	—	36	STAIGER 1954
		33	3	30	
		31	5	26	
		30	6	24	
		26	10	16	
Orthoptera	*Anaxipha pallidula*	19 (XO)	4	14	OHMACHI and UESHIMA 1957
		18	5	12	
		17	6	10	
	Ameles heldreichi	29 (XO)	—	28	WAHRMAN 1954
		28	1	26	
		27	2	24	
	Moraba scurra	17 (XO)	—	17	WHITE 1956
		16	1	15	
		15	2	13	
	Moraba viatica	19 (XO)	2	17	WHITE, CARSON, and CHENEY 1964
		17	4	13	
	Moraba virgo (parthenogenetic)	15 (XX, ♀)	4L + 1s	8s	WHITE, CHENEY, and KEY 1963
		15	3L + 2s	6s + 2L	
	Oedaleonotus enigma	22 (neo XY)	—	20	HEWITT and SCHROETER 1968
		21	1	18	
		20	2	16	
Coleoptera	*Chilocorus stigma*[1]	25 (X_1X_2Y)	22	—	SMITH 1959
		24	21	—	
		23	20	—	
		22	19	—	
		21	18	—	
		20	17	—	
		19	16	—	
	Pissodes terminalis	32 (XY)	2	28	MANNA and SMITH 1959
		30	4	24	
		29	5	22	
	Pissodes approximatus and *canadensis*	34 (XY)	—	32	
		33	1	30	
		32	2	28	
		31	3	26	
		30	4	24	
	Exochomus uropygialis	18 (XY)	—	16	SMITH 1965
		16	2	12	

Group	Species	2n	Autosomes		Reference
			V's	Rods	
Mammalia	*Gerbillus pyramidum*				
	(a) Algeria	40	38	2	WAHRMAN and
	(b) Israel-coastal plain	52	22–24	28–30	ZAHAVI 1955
	(c) Israel-Negev	66	8–12	44–58	
	(d) ♂ F_1 hybrid				
	(66 ♀ × 40 ♂)	53	24	29	
	Mus triton	22 (XY)	10	10	MATTHEY 1963
		21	11	8	
		20	12	6	
	Sus scrofa				
	(a) wild	37	25	10	MCFEE, BANNER,
	(b) domestic	36	26	8	and RARY 1966
Angiospermae	*Alisma plantago-*	14	10	4	CLAVIER (in
	aquatica	12	12	—	DARLINGTON 1963)
	Lycoris aurea	14	8	6	DARLINGTON 1963
		13	9	4	
		12	10	2	
	Northoscordum fragrans	19	13	6	LEVAN and
		18	14	4	EMSWELLER 1938
		16	16	—	

[1] The sequence in *Chilocorus stigma* is unusual in that it involves replacement of pairs of non-homologous metacentrics by single superficially similar metacentrics. This is accompanied by the elimination of heterochromatic arms or their conversion into B-chromosomes (see Fig. 76).

iii) Supernumerary Segments

A distinctive method of altering the morphology of a chromosome complement is through the addition of supernumerary segments to existing members of the karyotype. Such additions may involve either only one or else both homologues so that three distinct and distinguishable chromosome types can be found (Fig. 45). The origin of such supernumerary segments has rarely been adequately defined though they are known to occur in a number of orthopteroid insects (WHITE 1954). Two principal modes of origin suggest themselves—direct duplication or else translocation from supernumerary chromosomes either by non-reciprocal insertion or else reciprocal replacement. And both interstitial (NUR 1961) and terminal (JOHN and HEWITT 1966) supernumerary segments are known. A unique case has recently been described in the guinea pig *Cavia porcellus*. Here the largest autosome is normally acrocentric but both duplication and deletion of the short arms are known though as yet only in the heterozygous state (COHEN and PINSKY 1966). The authors suggest that the polymorphism arose via a translocation of the short arm between the two homologues of this chromosome element (Fig. 46). The net result is the production of states which are monosomic and trisomic respectively for the short arm of chromosome 1. Significantly the short arms of the chromosome are late replicating (see

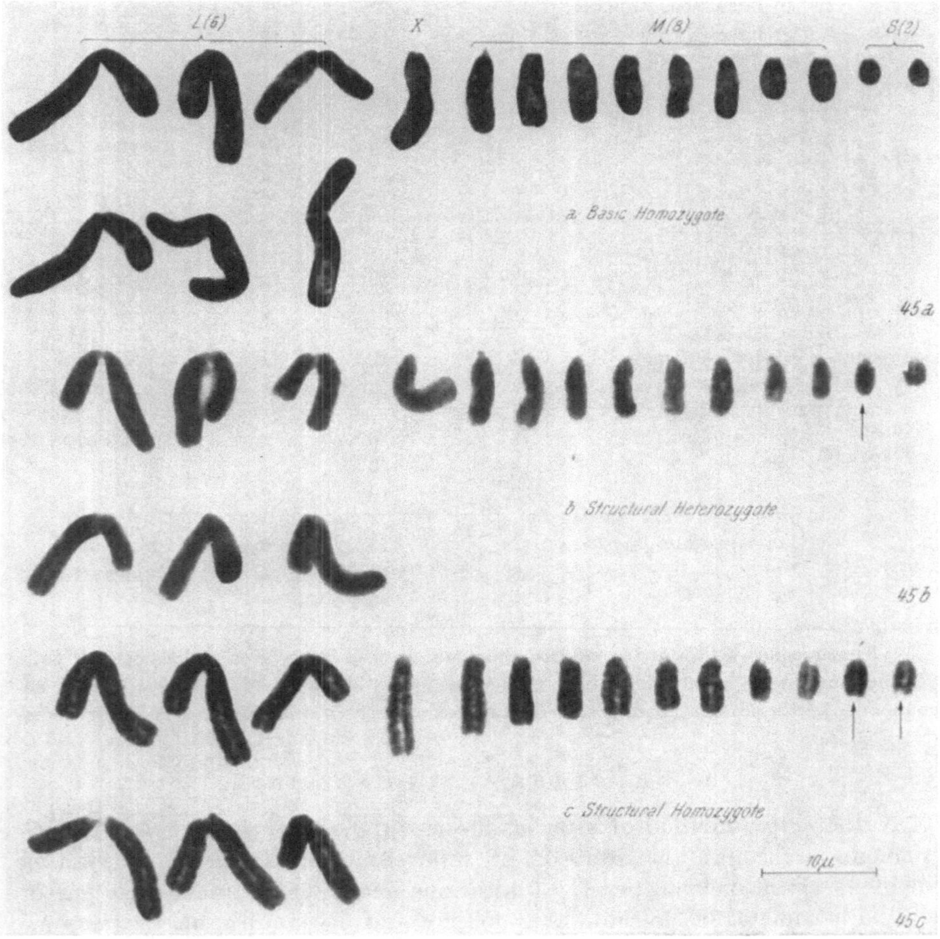

Fig. 45. Polymorphism for the occurrence of supernumerary segments (arrows) in the S_8 chromosome of *Chorthippus parallelus*.

pg. 69). It is worth recalling that in *Cavia cobaya* the heteromorphism in the nucleolar organising chromosome (see pg. 21) is remarkably similar to the trisomic state developed in *Cavia porcellus*.

iv) Compound Polymorphisms

The 5 chromosomes in the gametic set of *Trillium kamtschaticum* are designated as A–E in decreasing order of length. A complex and compound pattern of polymorphism has been found in these chromosomes by using the sequences of differential segments revealed by cold shock. This pattern can be shown to be constant for a given plant but varies from individual to individual. By this means 23 types of A, 14 of B, 9 of C, and 8 each of D and E have been distinguished (HAGA and KURABAYASHI 1953, 1954). These have originated by structural rearrangements and some, at least, of these have been defined (Fig. 47).

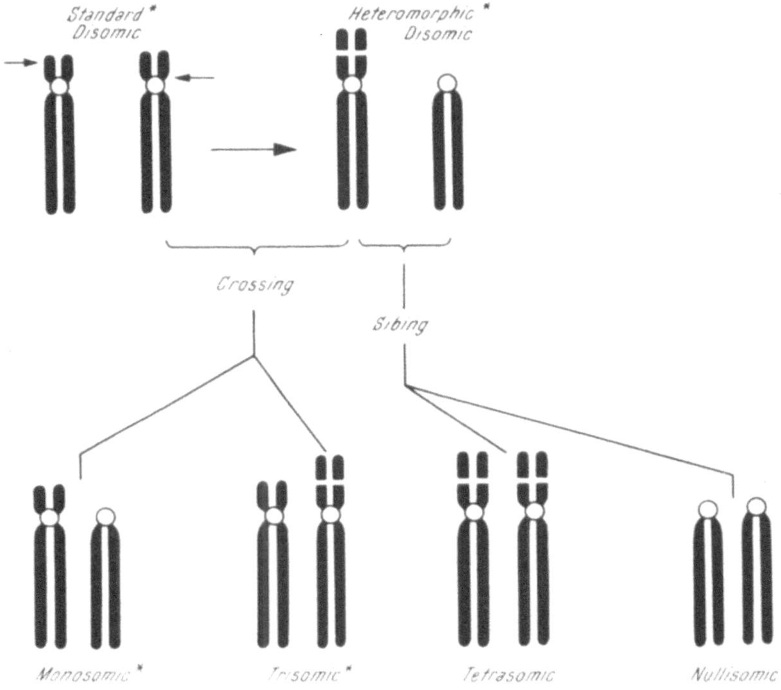

Fig. 46. Polymorphism for the presence of a short arm duplication in the guinea pig *Cavia porcellus* (based on COHEN and PINSKY 1966). The nature of the secondary constriction in the duplicate arm remains to be clarified. The asterisks refer to observed chromosome types.

A second instance of striking inter-plant variation has been found in *Anthoxanthum odoratum* (JONES 1964). Individuals differ both in the types of chromosomes present and in their relative frequencies. Indeed it is difficult to find two plants with identical karyotype. Variability is seen most readily in the SAT chromosomes and it appears that secondary constrictions may be transferred from one chromosome to another. Variation occurs also in the number of metacentric elements present in the karyotype. JONES is of the opinion that this species has originated from diploids whose chromosome sets differed in morphology and that the range of chromosome types is the consequence of recombination between the sets. Extensive structural differentiation of the chromosome complement is found also in different local populations of species belonging to the genus *Holocarpa* (CLAUSEN 1951).

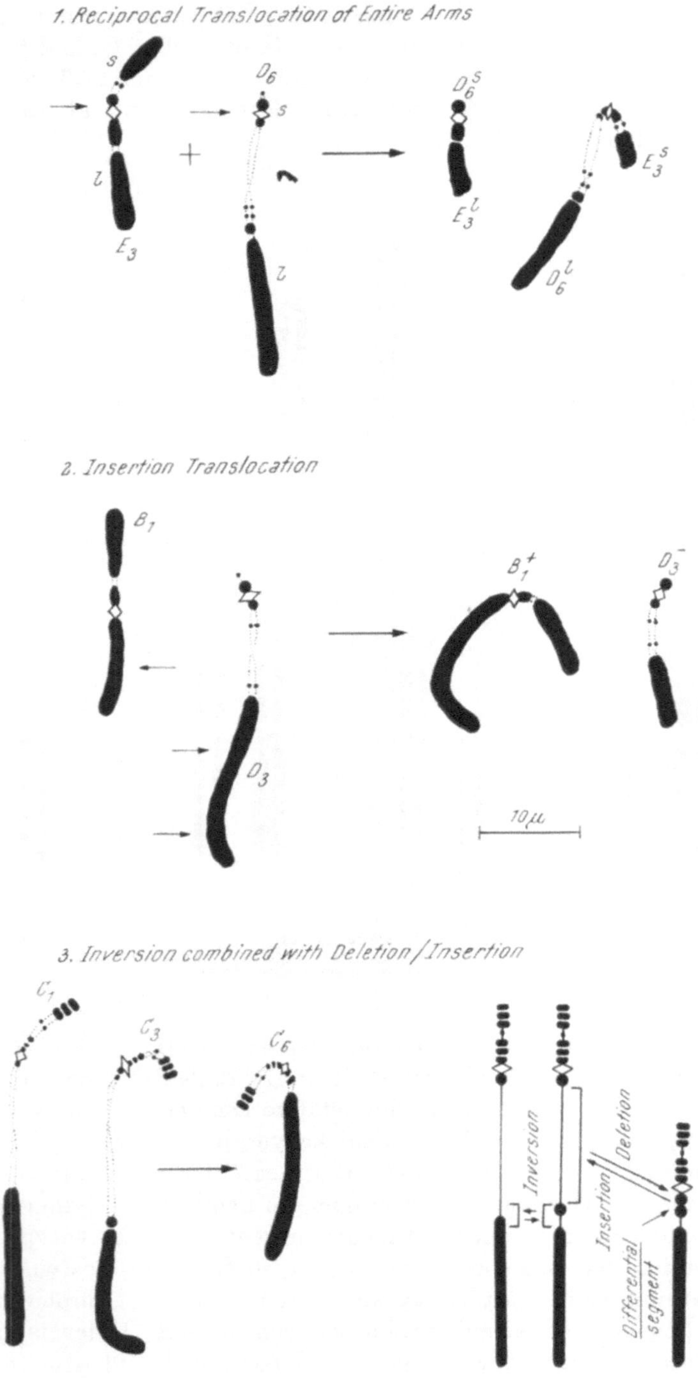

Fig. 47. Compound polymorphism for differential cold-induced heterochromatic segments in *Trillium kamtschaticum* (after Haga and Kurabayashi 1954).

Equivalent situations exist in animals and one of the most striking cases known occurs in the bark weevils *Pissodes approximatus* and *P. canadensis.* In both sexes of both species, the diploid number, although constant for the individual, ranges from 30 to 34. The numerical differences between individuals are due to two of the autosomes, designated as A and B, occurring either as V- or J- or rod-type elements (Fig. 48). The degree of polymorphism is enhanced by the incorporation of a pericentric inversion in

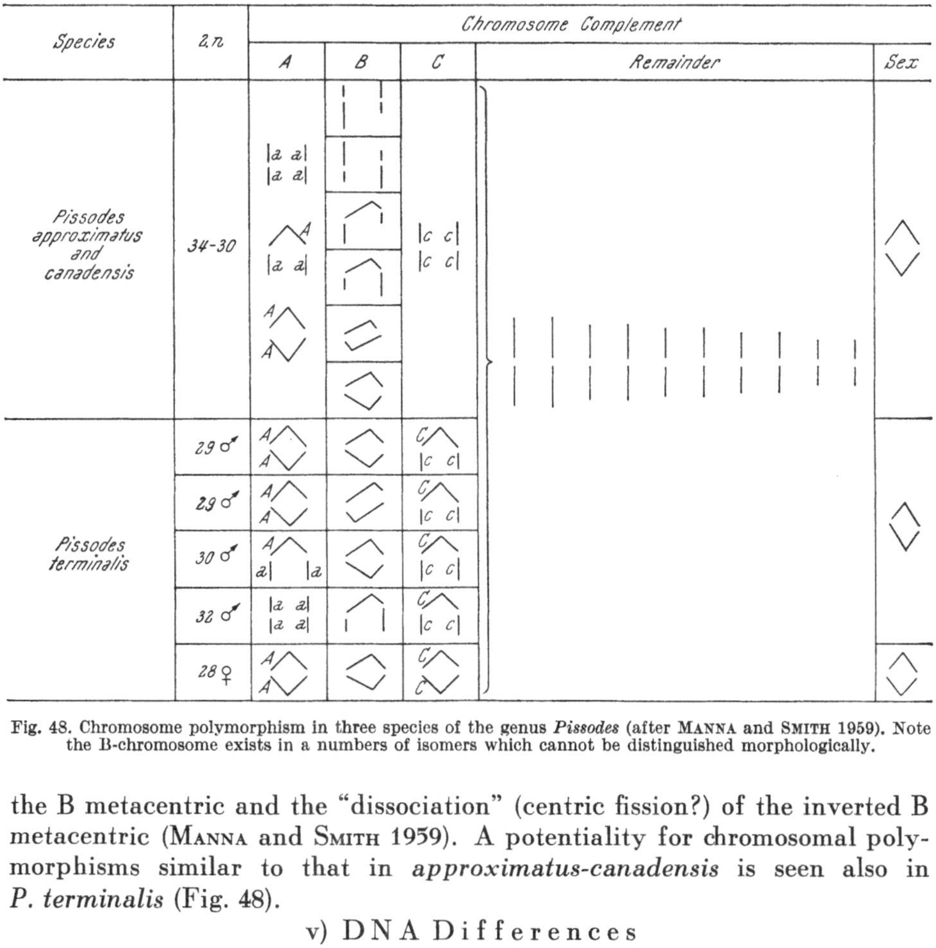

Fig. 48. Chromosome polymorphism in three species of the genus *Pissodes* (after MANNA and SMITH 1959). Note the B-chromosome exists in a numbers of isomers which cannot be distinguished morphologically.

the B metacentric and the "dissociation" (centric fission?) of the inverted B metacentric (MANNA and SMITH 1959). A potentiality for chromosomal polymorphisms similar to that in *approximatus-canadensis* is seen also in *P. terminalis* (Fig. 48).

v) DNA Differences

Chironomus thummi thummi and *Ch. thummi piger* differ in the DNA content of their spermatocytes, there being about 27% more DNA in *thummi thummi* (KEYL 1965, 1966). Studies on the polytene chromosomes of the two sub-species show how this DNA difference is distributed along the chromosome. Numerous bands in the mid region of the chromosomes have significantly higher DNA levels in *thummi thummi* than the homologous bands in *thummi piger.* Especially large differences are found in the kinetochore regions and in hybrids these regions are always unpaired thereby facilitating the estimation of the relative DNA content of homologous bands. Particular bands of *thummi thummi* contain 2, 4, 8, or 16 times more DNA than in the homologous bands of *thummi piger*. For certain loci two extinction ratios were found so that in particular bands within the one subspecies *thummi thummi* there may be a number of quantitative levels of DNA. This situation differs from the usual tandem duplication in that no new bands occur concurrently with the increases of DNA content and bands with higher DNA levels have a morphologically uniform appearance.

Similar variation in single bands has been found in natural populations of *Acricotopus* (PANITZ 1965). Most of the regions involved neither engage in puff production nor are they involved in puffs initiated by other bands. In two cases, however, (11/0-17 and 1/D-9) the presence of the high DNA mutant band in heterozygotes resulted in a reduction of both penetrance and expressivity as measured by the incidence and degree of puffing respectively.

3. Allosomal Polymorphism

In 1891 HENKING observed what he referred to as "a peculiar chromatin element" in the hemipteran *Pyrrhocoris*. This "element" lagged on the second division spindle after dividing at first and then passed undivided to one the two second division spindle poles. In consequence two equal classes of sperm were produced which could be distinguished by the presence or absence of the structure in question. HENKING had no clear appreciation of the nature of this "element" and doubtless it was for this reason that he labelled it "X"—the unknown! The first indication that this X-element was a chromosome came from MCCLUNG (1902) who emphasised the parallel between the two equally frequent classes of sperm differentiated by the presence or absence of the "X"-element and the 1 : 1 frequency of males and females.

We now recognise that what HENKING was dealing with was a system of male digamety involving an XO-XX sex-chromosome difference. Here the female complement includes two X-chromosomes and the male only one. The total number of chromosomes is thus even in the female but odd in the male and the single X-chromosome of the male must alternate between the two sexes in successive generations.

This simple XO♂, XX♀ type of sex chromosome system is now known in many animals (Table 37) though the size of the X varies widely in different species ranging from the largest (Fig. 49) to the smallest of the chromosomes in the complement.

A large number of animal species are, however, characterised by a system where, although the number of chromosomes is even and the same in both sexes, the members of the complement can be arranged in homomorphic pairs only in the homozygous sex. In the heterozygous, heterogametic sex one pair is asymmetrical or heteromorphic, most commonly consisting of a larger X and a smaller Y chromosome. Indeed the Y may be so small as to appear vestigial as in the case of the X_y coleopterans (Fig. 50).

Table 37. *Systems of Simple Male Digamety in Animals.*

System	Group	Species	Reference
XX/XY	*Acanthocephala*	*Acanthocephalus ranae*	JOHN 1957
	Ostracoda	*Cyclocypris laevis*	DIETZ 1958
	Chilopoda	*Thereuonema clunifera* and *hilgendorfi*	OGAWA 1962
	Acari	*Ornithodoros gurneyi*	OLIVER 1966
	Dermaptera	*Apterygida albipennis, Forficula auricularia* and *scudderi, Labia minor, Labidura bidens* and *riparia*	See Table 48
	Plecoptera	*Perla immarginata*	NAKAHARA 1919
	Aphaniptera	*Ctenocephalus canis*	BAYREUTHER 1954
	Neuroptera	Many species	See WHITE 1954
	Coleoptera	Many species	See SMITH 1953, 1960
	Diptera	Most species	See WHITE 1954
	Hemiptera —Heteroptera	Most species	See WHITE 1954
	Teleostei	*Bathylagus wesethi*	CHEN and EBELING 1966
	Reptilia	*Anolis conspersus* *Cupriguanus achalensis*	GORMAN and ATKINS 1966 GORMAN, ATKINS, and HOLZINGER 1967
	Mammalia	Most species	See WHITE 1954
XX/XO	*Nematoda*	*Ancyracanthus cysti* *Oswaldocruzia filiformis*	MÜLSOW 1911 JOHN 1957
	Ostracoda	*Cyclocypris globosa*	DIETZ 1958
	Isopoda	*Tecticeps japonicus*	NIIYAMA 1956
	Araneida	Several species	See MAKINO 1951
	Myriapoda	*Scolopendra cingulata, Scutigera forceps*	OGAWA 1950, 1952
	Collembola	*Entomobryidae, Sminthuridae*	NUNEZ 1962
	Orthoptera	Many species	See WHITE 1951
	Mecoptera	*Bittacus italicus* *Panorpa cognata, communis* and *germanica*	MATTHEY 1950 ULLERICH 1961

Table 37 (Continued)

System	Group	Species	Reference
XX/XO	*Plecoptera*	*Acroneuria jezoensis*, *Perla bipunctata* and *maxima*	ITOH 1933 MATTHEY and AUBERT 1947
	Odonata	Most species	OKSALA 1939–1945
	Coleoptera	Many species	See SMITH 1953, 1960
	Pscoptera	*Cerastipsocus venosus*	BORING 1913
	Embioptera	*Monotylota ramburi* *Oligotoma japonica*	LE CALVEZ 1949 KICHIJIO 1942
	Aphaniptera	*Leptopsylla segnis*	BAYREUTHER 1954
	Diptera	*Drosophila annulimana*, *longala*, *mercatorium* and *orbospiracula*; *Spathulina (Tephritis) arnicae*	PATTERSON and STONE 1952 KEUNEKE 1924
	Hemiptera (i) *Homoptera*	*Llaveia*, *Llaveiella*, *Nautococcus*, *Protortonia*, *Puto*	HUGHES-SCHRADER 1951
	(ii) *Heteroptera*	*Alydus*, *Anasa*, *Archimerus calcarator*, *Pachylis gigas*, *Protenor* and *Pyrrochoris*	WILSON 1905–1912
	Mammalia — *Monotremata*	*Tachyglossus aculeatus*	BICK and JACKSON 1967

Again the sex-chromosomes range in size from the smallest to the largest members of the complement. Thus in the beetles *Alagoasa extrema* and *Walterianella venusta* (VIRKKI 1964) and in the chilopod *Thereuonema hilgendorfi* (OGAWA 1962) the X and Y are of giant size compared to the autosomes (Fig. 51).

As a group, however, the XX/XY systems are heterogeneous in the sense that while it is the basic condition in some species it is a derived one in others. These latter cases, which have arisen secondarily from XX/XO systems, are spoken of as neo-XY mechanisms (Figs. 43 and 44). They arise as a result of an interchange between the X and one member of an autosomal pair. In consequence the unmodified autosome becomes the new Y while the X^A interchange product forms the neo X. The reciprocal product of interchange (A^X) must be lost so, clearly, the XO→XY progression is likely only when acro- or telo-centric chromosomes are involved. The net result is the inclusion of the bulk of an autosomal linkage group into a sex chromosome and thus the reduction of the number of linkage groups by one (Table 38).

In *Pales ferruginea* the X-chromosome is found in two forms which differ in length—X_l and X_k. The ratio l : k is in the order of 1.4 : 1 (Fig. 52).

The two types show normal mendelian behaviour (Table 39) and are viable in all combinations (ULLERICH, BAUER, and DIETZ 1964). X-chromosomes of different length have also been observed in *Tipula paludosa* and *T. oleracea*. In *Phryne cincta* too different types of X-chromosomes occur in natural

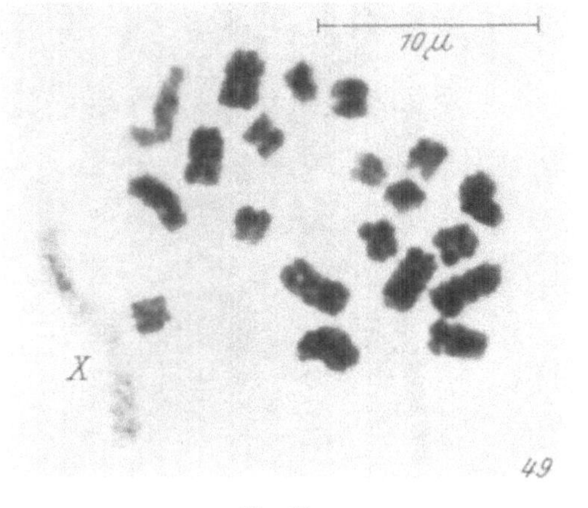

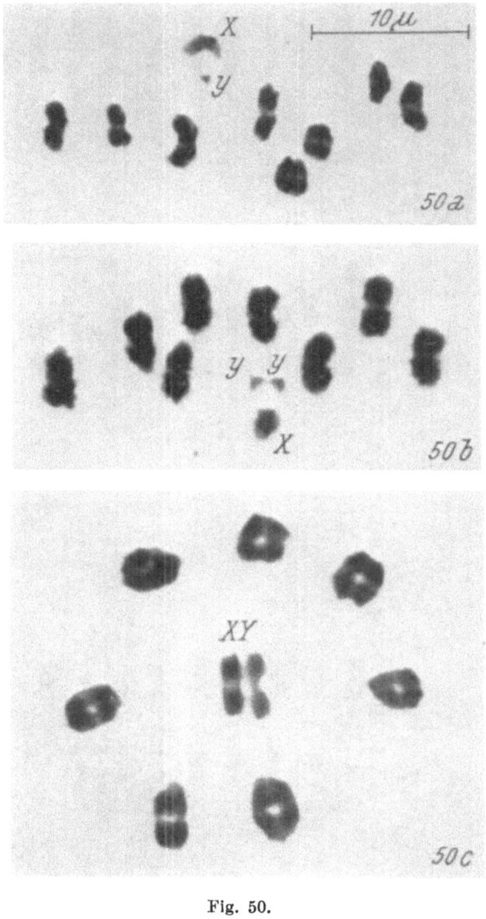

Fig. 49. Fig. 50.

Fig. 49. Mitotic prophase of the cricket *Achaeta domestica* to show the large and negatively heteropycnotic X-chromosome (♂ 2n = 21 = 20 + X).

Fig. 50. Sex-chromosome types at first meiotic metaphase in *Dermestes maculatus* (Figs. 50*a* and *b*) and *Oncopeltus fasciatus* (Fig. 50*c*). Fig. 50*a* shows the standard condition in *D. maculatus* (♂ 2n = 18 = 8 II + Xy) while Fig. 50*b* is taken from an individual with a supernumerary y-chromosome (♂ 2n = 19 = 8 II + Xyy). In *Oncopeltus* (Fig. 50*c*) the complement includes seven pairs of autosomes and a pair of hereditary sex univalents (♂ 2n = 16 = 7 II + XY) and all the chromosomes in this species have a non-localised kinetic system.

populations as distinct races (WOLF 1956) and here the long (L) and short (K) forms have a ratio of 2 : 1. By crossing KK♀ with LY♂ or LL♀ with KY♂ it is possible to produce LK♀ heterozygotes and a study of the pairing relationships of the two X-chromosomes in polytene nuclei shows that the two X's differ not only in length but also in the order of their band sequences (Fig. 53).

Cases occur also where the X is represented not by one but by two or more members. Numbers ranging from 2 to 8 have been observed (Table 40).

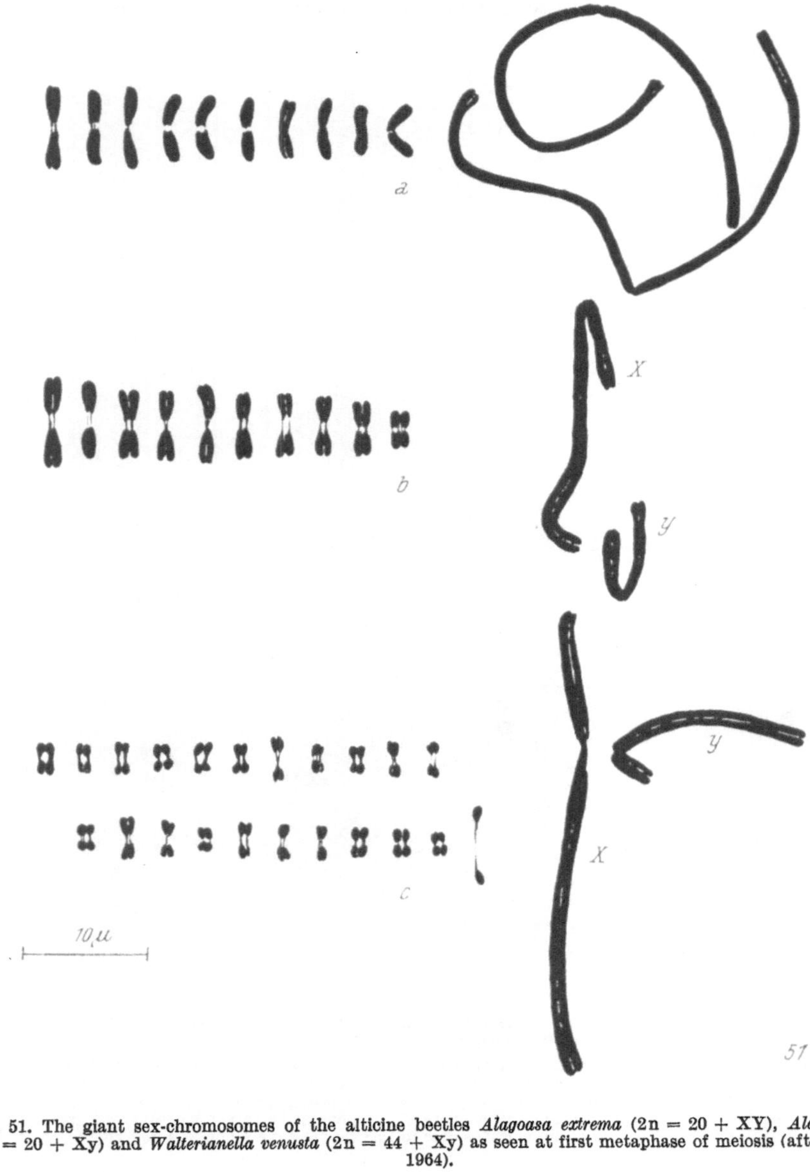

Fig. 51. The giant sex-chromosomes of the alticine beetles *Alagoasa extrema* (2n = 20 + XY), *Alagoasa sp.* (2n = 20 + Xy) and *Walterianella venusta* (2n = 44 + Xy) as seen at first metaphase of meiosis (after VIRKKI 1964).

Whatever their number the X group is hemizygous in the male sex but is represented twice in the female. Alternatively the Y component may be multiple while the X remains as a single element. Even more remarkable conditions are found among those species where a compound X-system is coupled with a multiple Y mechanism. A unique sex-chromosome system has been described in the indian mongoose (*Herpestes auropunctatus*). Here males have 35 and females 36 chromosomes in bone marrow cells, the males having one and the females two X-chromosomes. At meiosis in the male, however, 16. II + 1. III are formed, the multiple of 3 consisting of a single X and one "pair" of autosomes (FREDGA 1964). FREDGA suggests that the

Table 38. *Neo Sex-Chromosome Systems.*

System	Group	Species	Reference
Neo XY	*Orthoptera*	Numerous species	See WHITE 1957 and SAEZ 1963
	Coleoptera	Many species	See SMITH 1953, 1960
Neo X_1X_2Y	*Orthoptera*	Many mantids Several morabines *Paratylotropidia brunneri* and *morsei*	See WHITE 1957
	Diptera	*Drosophila miranda*	PATTERSON and STONE 1952
	Marsupialia	*Lagorchestes conspicillatus*	MARTIN and HAYMAN 1966
Neo XY_1Y_2	*Diptera*	*Drosophila americana americana*	PATTERSON and STONE 1952
	Marsupialia	*Potorous tridactylus* *Protemnodon bicolor*	SHARMAN and BARBER 1952 SHARMAN 1961

Table 39. *The Inheritance of X-chromosome Types in Pales ferruginea.* (Data of ULLERICH, BAUER, and DIETZ 1964.)

Parental Types (♀ × ♂)	Constitution of Offspring				
	Female			Male	
	X_lX_l	X_lX_k	X_kX_k	X_lY	X_kY
1. $X_lX_k \times X_lY$	15	17	—	18	14
2. $X_kX_k \times X_kY$	—	—	23	—	16
3. $X_lX_k \times X_kY$	—	9	12	8	8

system is a derivative of an XY condition in which that part of the Y which associated terminally with the X at meiosis has become translocated to an autosomal segment the remaining centric region of the Y having been lost (Fig. 54). Consequently at meiosis the two "autosomes" (A^a and A^y) form a bivalent and the X associates terminally with the A^y to form a multiple of three. If this interpretation of the sex multiple is correct then, clearly, it should really be regarded as a neo X_1 X_2 Y association ($\equiv XA^a A^y$ respectively) and one which presents segregational problems.

In the *Lepidoptera* and *Aves* and in certain copepods, isopods, fish, amphibia and reptiles the female is the heterozygous sex. In these circumstances instead of simply reversing the symbols employed in male digamety it is convention to use a distinct terminology substituting Z for X and W for Y. In these terms one can distinguish ZO/ZZ and ZW/ZZ systems (Table 41).

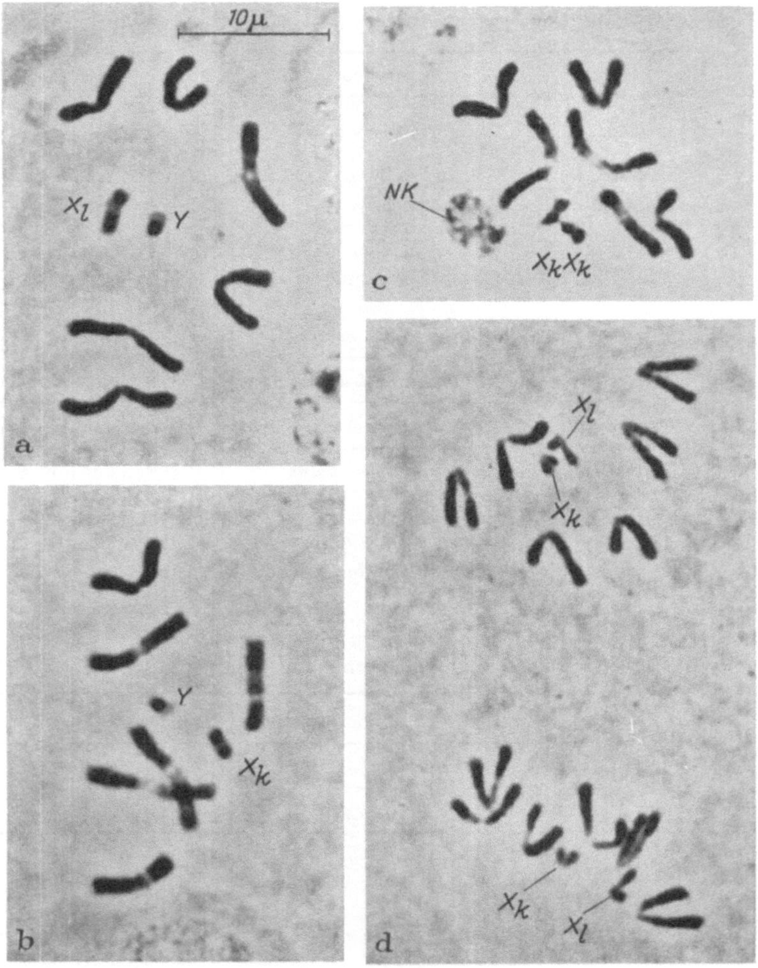

Fig. 52. Long (X_l) and short (X_k) variants of the X-chromosome in *Pales ferruginea* as seen at spermatogonial metaphase (Figs. 52*a* and *b*) and at metaphase (Fig. 52*c*) and anaphase (Fig. 52*d*) of secondary oogonia (after Ullerich, Bauer, and Dietz 1964).

While approximately three-quarters of the families of flowering plants include dioecious members in only about 5% of species are the sexes separate. Claims for the existence of morphologically distinguishable sex-chromosomes are numerous. But most of the species concerned have small chromosomes, many of the early claims are unconfirmed and cannot, therefore, be regarded as established. Indeed, even in extensively studied cases

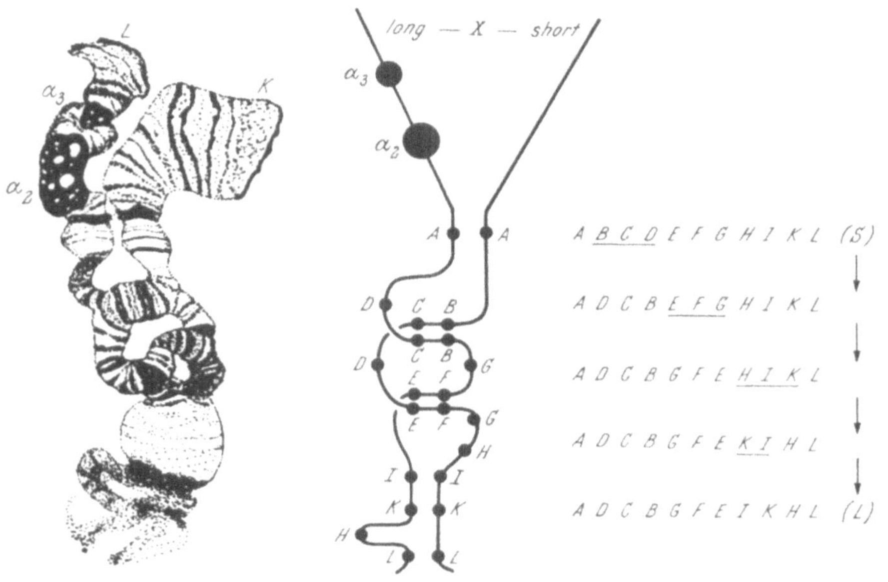

Fig. 53. The pairing homologies of the short (K) and long (L) forms of the X-chromosome of *Phryne cincta* as seen in salivary gland nuclei (after WOLF 1956).

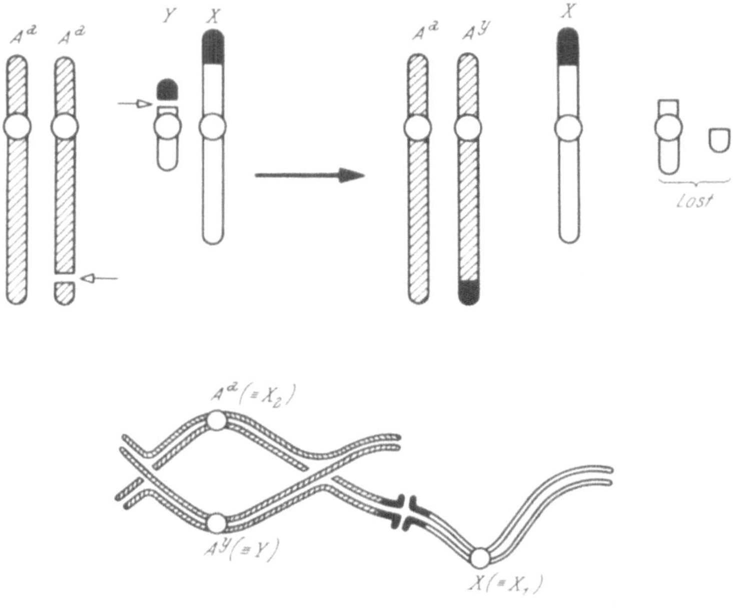

Fig. 54. The proposed origin of the sex multiple in the indian mongoose *Herpestes auropunctatus*. The original pairing segment between the X and the Y is shown solid (after FREDGA).

such as *Spinacia oleracea* and *Dioscorea* (MARTIN 1966) agreement has not been reached. For these reasons the number of accepted examples in angio-

Table 40. *Multiple Sex Chromosome Systems in Animals.*

System	Group	Species	Reference
X_nX_n/X_nO	*Nematoda*	*Belascaris triquetra, Ganguleterakis spumosa* and *Rhabdias fülleborni* (X_1X_2) *Ascaris lumbricoides* (X_1-X_5) *Toxascaris canis* (X_1-X_6) *Contracaecum incurvum* (X_1-X_8)	GOODRICH 1914, 1916 WALTON 1916, 1924
	Ostracoda	*Notodromonas monacha* (X_1X_2) *Platycypris baueri, Physocypria kliei* and *Scottia browniana* (X_1-X_3)	DIETZ 1958
	Pseudo-scorpionida	*Chelanops cyrneus* and *Obisium muscorum* (X_1X_2)	
	Araneida	Most spiders (X_1X_2 or X_1-X_3)	REVELL 1947, SUZUKI 1952
	Thysanura	*Thermobia domestica* (X_1X_2)	PERROT 1933
	Plecoptera	*Perla cephalotes* (X_1X_2) *Perla jurassica* and *microcephala* (X_1-X_3)	MATTHEY and AUBERT 194[illegible]
	Hemiptera (i) *Homoptera*	*Phylloxera caryaecaulis* (X_1X_2) *Euceraphis betulae* (X_1-X_4) *Matsucoccus gallicola* (X_1-X_6)	MORGAN 1912 SHINJI 1931 HUGHES-SCHRADER 1948
	(ii) *Heteroptera*	*Dysdercus koenigii, Haploprocta sulicornis, Holopterma alata* and *Syromastes marginatus* (X_1X_2) *Coreus marginatus* (X_1-X_3)	See MANNA 1958, 1962
X_nX_n/X_nY	*Ostracoda*	*Cypris compacta, dietzi, fodiens* and *whitei* (X_1-X_3Y) *Cypria exsculpta* and *ophthalmica* (X_1-X_4Y) *Cypria exsculpta* (X_1-X_5Y) *Heterocypris incongruens* (X_1-X_6Y)	DIETZ 1958
	Acari	*Amblyomma limbatum* and *moreliae* (X_1X_2Y)	OLIVER 1965
	Coleoptera	*Blaps lethifera* and *lusitanica, Chilocorus stigma,* and *Cicindela repanda, scutellaris, sexguttata* and *tranquebarica* (X_1X_2Y) *Blaps mortisaga* and *mucronata, Cicindela campestris, hybrida, lunulata* and *silvicola* (X_1-X_3Y) *Blaps gigas* (X_1-X_4Y)	GUÉNIN 1948–1952 SMITH 1959 SMITH and EDGAR 1954

System	Group	Species	Reference
X_nX_n/X_nY	*Dermaptera*	*Anisolabis annulipes, marginalis* and *maritima* (X_1X_2Y) *Prolabia arachidis* (X_1-X_3Y)	Bauer 1947, Morgan 1928, Schrader 1941, Sugiyama 1933
	Aphaniptera	*Nosopsylla fasciata* and *Xenopsylla cheopis* (X_1X_2Y)	Bayreuther 1958
	Mecoptera	*Boreus brumalis* (X_1X_2Y)	Cooper 1951
	Diptera	*Anastrepha serpentina, Rhagoletis striatella* and *Hylemya fugax* (X_1X_2Y)	Boyes 1952, Bush 1962, 1966
	Hemiptera (i) *Homoptera*	*Conorhinus sanguisugus* (X_1X_2Y) and *Sinea diadema* (X_1-X_3Y)	Wilson 1911
	(ii) *Heteroptera*	Many species	See Table 68 in John and Lewis 1965 and Ueshima 1966
	Reptilia	*Anolis bimaculatus, biporcatus, fereus, gingivinus, leachi* and *marmoratus* (X_1X_2Y) *Sceloporus jarrovi* and *poinsetti* (X_1X_2Y) *Polychrus marmoratus* (X_1X_2Y)	Gorman and Atkins 1966 Cole and Lowe 1967 Gorman, Atkins, and Holzinger 1967
	Mammalia	*Mus min. minutoides* ssp. 3 (X_1X_2Y)	Matthey 1965
XX/XY_n	*Ostracoda*	*Cyclocypris ovum* (XY_{3-7})	Dietz 1958
	Orthoptera	*Eneoptera surinamensis* (XY_1Y_2)	Piza 1946
	Mammalia	*Gerbillus gerbillus* and *Sorex araneus* (XY_1Y_2)	Wahrman and Zahavi 1955 Ford *et al.* 1957
X_nX_n/X_nY_n	*Myriapoda*	*Otocryptops sexspinosus* (X_4Y_5)	Ogawa 1954
	Coleoptera	*Blaps cribosa* ($X_{12}Y_6$)	Guénin 1953 and Wahrman unpublished (see John and Lewis 1965)
	Diptera	*Drosophila prosaltans* II ($X_1X_2Y_1Y_2$)	Patterson and Stone 1952

sperms are few and are given in Table 42. Dioecism is common among bryophytes, typifying about a third of the species. Here, of course, sex is exhibited in the haplophase (♀ $= n =$ X, ♂ $= n =$ Y and sporophyte $= 2n =$ XY). Among the liverworts XY and X_1X_2Y systems have been described though the observations of different investigators are not always consistent (see Berrie 1960 and Lewis 1961).

Table 41. *Systems of Female Digamety in Animals.*

System	Group	Species	Reference
ZO/ZZ	*Copepoda*	*Ectocyclops strenzkei* *Eucyclops serrulatus*	BEERMANN 1954 RÜSCH 1960
	Lepidoptera	*Phragmatobia fuliginosa* *Fumea casta* *Talaeporia tubulosa*	SEILER 1914, 1917, 1921
ZW/ZZ	*Trematoda*	*Schistosomatium douthitti*	SHORT 1957
	Copepoda	*Acanthocyclops vernalis* *Megacyclops viridis*	RÜSCH 1960
	Lepidoptera	Numerous species	See WHITE 1954
	Diptera	*Chrysotrypanea trifasciata* (+ 7 other unnamed species)	BUSH 1966
	*Amphibia**	*Discoglossus pictus*	MORESCALCI 1964
	Reptilia	*Bothrops jararaca*	BEÇAK, BEÇAK and NAZARETH 1962
		Vipera berus	KOBEL 1962
	Aves	*Melopsittacus undulatus* *Gallus domesticus* Probably all birds	ROTHFELS 1963 OWEN 1965
ZW_1W_2/ZZ	*Copepoda*	*Jaera marina*	STAIGER and BOCQUET 1954

* The case of *Xenopus laevis* has been excluded since MIKAMO and WITSCHI (1966) have given good grounds for discrediting the claims of WEILER and OHNO (1962) and MORESCALCHI (1963).

Table 42. *Sex Chromosome Systems in Angiosperms.*

System	Species	2n	Reference
XX/XY	*Humulus lupulus*	20	JACOBSEN 1957
	Melandrium album	24	WESTERGAARD 1940
	rubrum	24	LÖVE, D. 1944
	Rumex angiocarpus	14	LÖVE, A. 1943
	paucifolius	28	LÖVE, A. and SARKAR 1956
	tenuifolius	28	LÖVE, A. 1943
	acetosella	42	LÖVE, A. 1943
	graminifolius	56	LÖVE, A. 1943
XX/XY_1Y_2	*Humulus japonicus*	16/17	JACOBSEN 1957
	Rumex hastatulus	8/9	SMITH 1955
	acetosa	14/15	YAMAMOTO 1938
$X_1X_1X_2X_2/X_1X_2Y_1Y_2$	*Humulus lupulus*	20	ONO 1937

V. Inter-Specific Karyotype Variation

1. Differences in Number

In dealing with the question of numerical differences between the chromosome complements of different species two cautions are necessary. First, to speak of the number of chromosomes as a specific characteristic does not imply that this number is absolutely fixed. In fact, as we have already seen, deviations from the typical number are often observed within a species and indeed even in different cells of the same individual. Secondly the history of cytology leaves us in doubt that counting chromosomes correctly is apparently not so simple an exercise as has sometimes been assumed. The degree of uncertainty in pioneer studies is especially clear in the case of man.

With these cautions in mind let us note that the chromosome number has been counted with varying degrees of accuracy in representatives of all the major groups of plants and animals. There is little to be gained from listing these counts since this has already been done in detail by DARLINGTON and WYLIE (1956) for plants and by MAKINO (1951) for animals. There are, however, several general points which are worth making:

(1) Although the chromosome number may vary between very wide limits (Table 43), a majority of species have haploid numbers between 6 and 20 and numbers above and below these limits are not common.

(2) The higher numbers are more common in plants than animals. This probably reflects the higher incidence of polyploidy in plants.

(3) Some groups show considerable variation in chromosome number (e.g., *Lepidoptera, Trichoptera,* Decapod Crustacea, Scorpions, Fishes and Urodeles among animals and the *Gramineae, Compositae* and *Ranunculaceae* among plants) while in others the complement shows considerable numerical stability (*Corixidae, Odonata, Diptera, Acrididae* and *Coleoptera* among animals and Gymnosperms among plants).

With these facts in mind let us now examine the methods by which decreases and increases in number may be achieved. We might suppose that increase in basic number would be easier to effect than a decrease since a gain of genetic material is usually less deleterious than a loss.

a) Mechanisms Leading to Decrease

Decreases in basic chromosome number can be initiated only by unequal reciprocal translocations. This involves the translocation of the greater part of one chromosome onto a non-homologous one. The latter, in exchange, contributes only a small segment. Provided therefore the small fragments can be lost there will be a net reduction of chromosome number by one for each translocation. For example in the plant genus *Haplopappus* the available evidence suggests that the basic chromosome number is $x = 4$ and that this has, on occasions, been reduced to 3 and 2. In fact, collections of *H. gracilis* from a single area in south-central Arizona gave plants with $2n = 4$, 5 and 6 in the same population (JACKSON 1965). The $2n = 5$ race

Table 43. *Modal and Extreme Haploid Chromosome Numbers in Plants and Animals.* The list is intended to be illustrative rather than exhaustive and deals only with some of the groups in which a sufficient number of counts exist to make a range selection feasible; numerical variation caused by B-chromosomes is disregarded.

Group		Haploid Numbers		
		Lowest	Modal	Highest
(1) Plants	1. *Bryophyta*			
	(A) *Hepaticae*			
	(1) *Marchantiales*	3	9	36
	(2) *Takakiales*	4	4	4
	(3) *Anthocerales*	4	5	10
	(4) *Jungermanniales*			
	(a) *Anacrogyne*	4	9	30
	(b) *Acrogyne*	8	9	36
	(5) *Sphaerocarpales*	8	8	9
	(B) *Musci*			
	(1) *Eubryales*	5	6	40
	(2) *Fissidentales*	5	16	21
	(3) *Hypnobryales*	5	10	24
	(4) *Isobryales*	6	11	22
	(5) *Polytrichales*	7	7	22
	(6) *Pottiales*	9	13	66
	(7) *Dicranales*	10	13	42
	(8) *Sphagnales*	21	21	42
	2. *Pteridophyta*			
	(1) *Lycopodiales*	9	9	78
	(2) *Filicales*	13	22	205
	(3) *Equisetales*	c. 108	c. 108	c. 108
	3. *Gymnospermae*			
	(1) *Ephedraceae*	7	7, 14	14
	(2) *Cycadaceae*	8	9	13
	(3) *Podocarpaceae*	10	19	20
	(4) *Cupressaceae*	11	11	22
	(5) *Pinaceae*	12	12	22
	4. *Angiospermae*			
	(A) *Dicotyledoneae*			
	(1) *Cruciferae*	4	7, 8	56
	(2) *Compositae*	4	9	72
	(3) *Papaveraceae*	5	6, 7	35
	(4) *Onagraceae*	5	7, 11	38
	(5) *Ranunculaceae*	5	8, 16	70
	(6) *Papilionaceae*	5	7, 8, 11, 24	c. 90
	(7) *Berberidaceae*	6	6, 14	28
	(8) *Myrtaceae*	6	11	44
	(9) *Umbelliferae*	6	11	48
	(10) *Polygonaceae*	7	10, 11	c. 100
	(11) *Caprifoliaceae*	8	9	27
	(12) *Ficoidaceae*	8	9, 18	36
	(13) *Ericaceae*	8	12, 26	72
	(14) *Oleaceae*	11	13, 23	69
	(15) *Magnoliaceae*	19	19, 57	57

Group		Haploid Numbers: Lowest	Modal	Highest
(1) Plants	(B) *Monocotyledoneae*			
	(1) *Liliaceae*	3	7, 8, 12	54
	(2) *Iridaceae*	3	10	90
	(3) *Gramineae*	5	7, 10, 14, 18, 20	90
	(4) *Araceae*	8	14	c. 70
	(5) *Zingiberaceae*	9	9, 24	c. 51
	(6) *Orchidaceae*	9	20	63
	(7) *Bromeliaceae*	16	25	c. 63
(2) Animals	1. *Platyhelminthes*			
	(A) *Turbellaria*			
	Rhabdocoela	2	2	10
	(B) *Trematoda*			
	Digenea	6	11	14
	2. *Mollusca*			
	Gastropoda			
	Pulmonata	6	29	31
	3. *Arthropoda*			
	(A) *Crustacea*			
	Decapoda	34	53, 58, 62	127
	(B) *Insecta*			
	(1) *Hemiptera*			
	(a) *Coccoidea*	2	2, 5	20
	(b) *Pentatomidae*	3	7	14
	(c) *Aphididae*	3	6, 7	20
	(d) *Miridae*	12	17	24
	(2) *Diptera*	2	4, 6	9
	(3) *Coleoptera*	4	9, 10, 11	29
	(4) *Hymenoptera*	5	8, 10	16
	(5) *Orthoptera*			
	(a) *Gryllidae*	5	10	15
	(b) *Acrididae*	6	12	12
	(c) *Manteidae*	8	14	20
	(d) *Tettigoniidae*	10	16	34
	(6) *Lepidoptera*	8	29, 30, 31	191
	(7) *Odonata*	9	13, 14	14
	4. *Chordata*			
	(A) *Pisces*			
	Teleostei	9	23, 24	52
	(B) *Amphibia*			
	(1) *Urodela*	9	12, 14, 28	32
	(2) *Anura*	11	11, 13	18
	(C) *Reptilia*			
	Lacertilia	11	19, 23	32
	(D) *Aves*	25	40	43
	(E) *Mammalia*			
	(1) *Marsupialia*	5	6, 11	14
	(2) *Placentalia*	7	21, 24	39

proved to be a hybrid between the 4 and 6 types and, at meiosis, regularly gave 1 II + 1 III (Fig. 55). The relationships of the chromosome complements in the 6 and 4 types, together with their pairing behaviour in the 2 n = 5 hybrid, implicated a process of aneuploid reduction by an unequal translocation (Fig. 56). In a like manner the pairing relationships of the chromosomes of *H. gracilis* (2 n = 4) and *H. ravenii* (2 n = 8) when brought

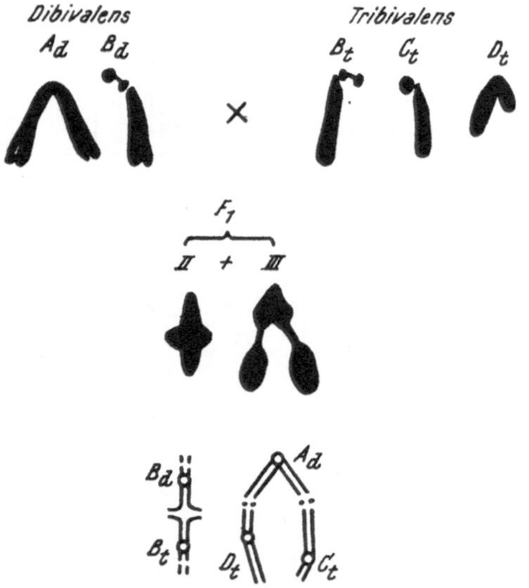

Fig. 55. The behaviour of the chromosomes in the F_1 hybrid between the two chromosome 'races' of *Haplopappus gracilis* (based on JACKSON 1965).

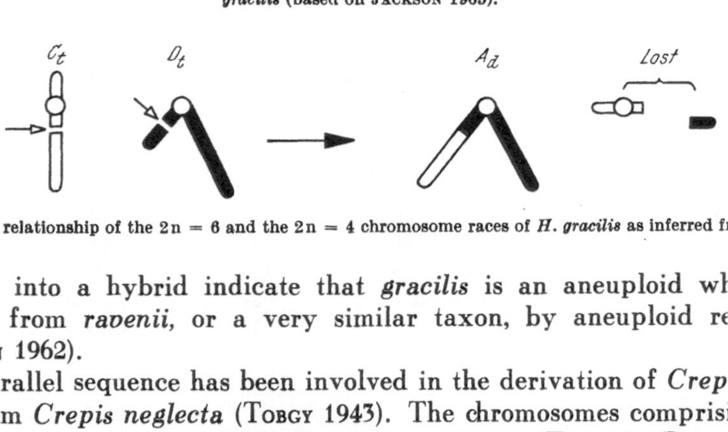

Fig. 56. The relationship of the 2n = 6 and the 2n = 4 chromosome races of *H. gracilis* as inferred from Fig. 55.

together into a hybrid indicate that *gracilis* is an aneuploid which has evolved from *ravenii,* or a very similar taxon, by aneuploid reduction (JACKSON 1962).

A parallel sequence has been involved in the derivation of *Crepis fulginosa* from *Crepis neglecta* (TOBGY 1943). The chromosomes comprising each complement are again individually distinguishable (Fig. 57). By analysing the meiotic pairing in hybrids between them TOGBY was able to show that two of the chromosomes of the haploid set in *neglecta* (B_N and C_N) are represented by only one chromosome (B_F) in *fulginosa,* the C-chromosome being absent from the complement of the latter species. Not all the material of the C-chromosome is, however, lacking in *fulginosa.* At meiosis in the F_1 hybrid the B and C chromosomes of *neglecta* pair with the B of *fulginosa* and in such a way as to indicate that these chromosomes differ with respect to an unequal and reciprocal translocation (Fig. 58). Other pairing configurations in this hybrid suggest that the parental species are differentiated in respect of several other rearrangements. Thus the A and D chromosomes of *fulginosa* differ from the comparable chromosomes of *neglecta* by one

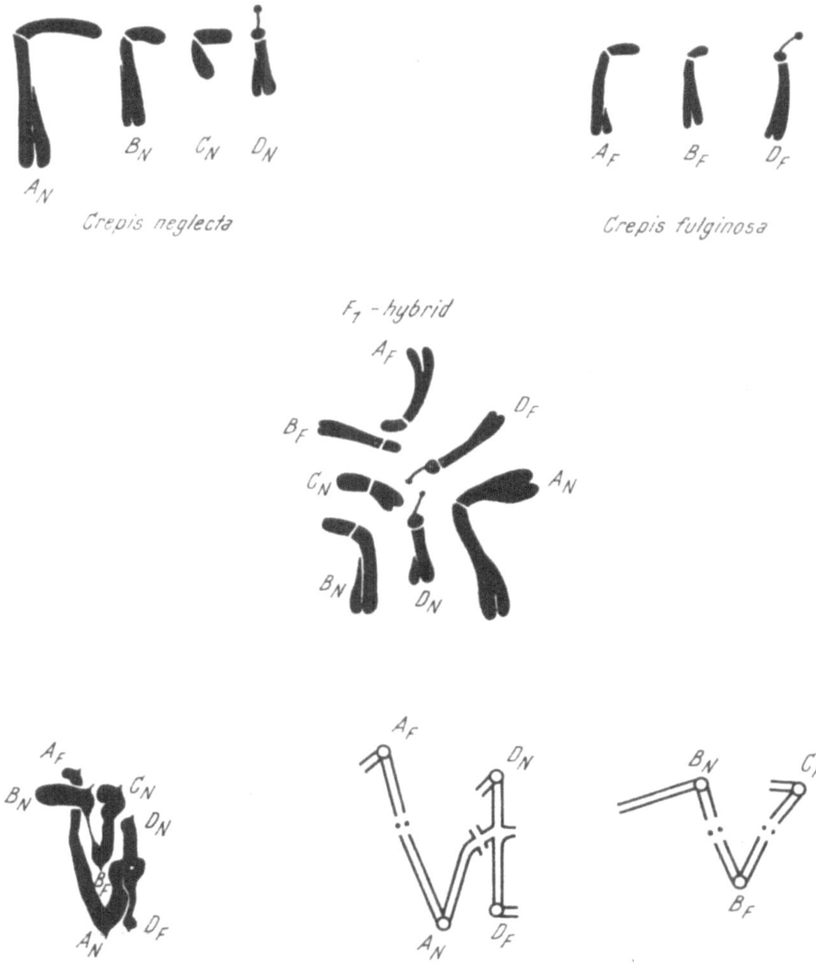

Fig. 57. The behaviour of the chromosomes in the F_1 hybrid between *Crepis neglecta* (2n = 8) and *C. fulginosa* (2n = 6).

unequal translocation. Moreover the long arms of the A, B and D in *fulginosa* may carry segments that are inverted relative to the homologous regions in *neglecta*. SHERMAN (1946) has shown that the 4-paired *Crepis kotschyana* was derived in a similar manner from a five-paired form related to *C. foetida*.

A comparable though more complex example of this same process of aneuploid reduction has recently been described by KYHOS (1965) in the composite genus *Chaenactis.* He studied three annual species which are morphologically very similar. Two of these are restricted in distribution and have $n = 5$ while the third is more widespread in California and has $n = 6$. Since $n = 6$ is the only number known in the relatively primitive perennial species of this genus, and since this same number also categorises the greatest array of distinctive species, KHYOS has argued that this must represent the basic number so that the $n = 5$ forms are secondarily derived.

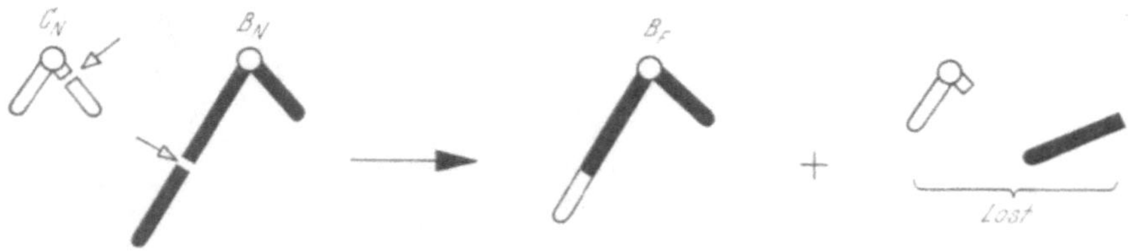

Fig. 58. The relationship of *Crepis neglecta* and *C. fulginosa* as inferred from Fig. 57.

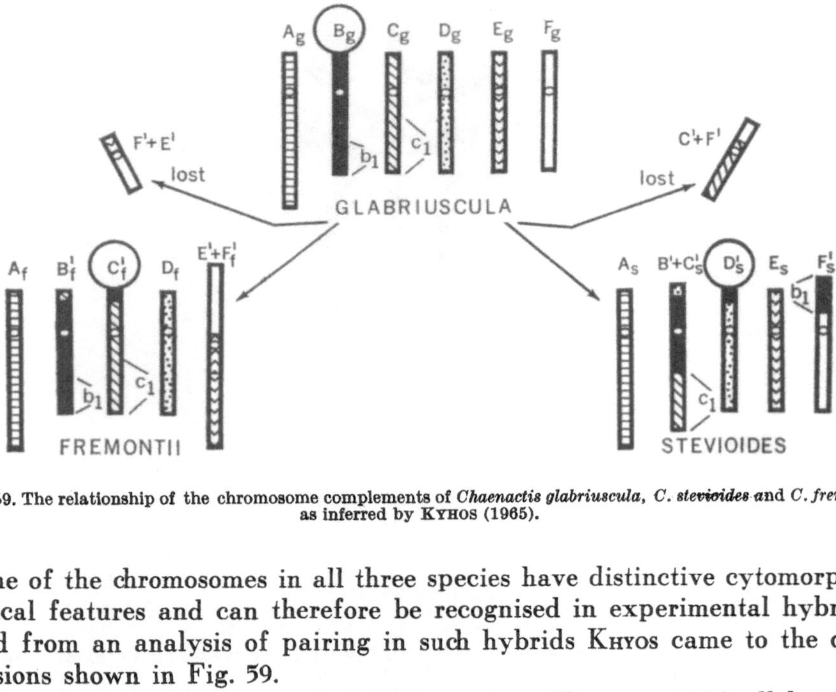

Fig. 59. The relationship of the chromosome complements of *Chaenactis glabriuscula*, *C. stevioides* and *C. fremontii* as inferred by KYHOS (1965).

Some of the chromosomes in all three species have distinctive cytomorphological features and can therefore be recognised in experimental hybrids. And from an analysis of pairing in such hybrids KHYOS came to the conclusions shown in Fig. 59.

Again in the liliaceous genus *Polygonatum* (THERMAN 1953) all but one of the species in the group *Alternifolia* have a haploid number of 10 or some multiple of this. The species *multiflorum* is unique in having a diploid count of 18. The hybrid between *multiflorum* and *officinale* has 19 chromosomes which at meiosis give rise to 8 II + 1 III. This relation has, again, most probably arisen through an unequal translocation between two chromosomes in the $2n = 20$ set followed by loss of the smaller of the resulting chromosomes.

Table 44. *Relationship Between V and Rod-type Chromosomes in Related Species.*

Group	Species	2n	Autosomes		Reference
			V's	Rods	
Orthoptera					
(i) *Cyrtacanth-acridinae*	*Dichroplus conspersus, elongatus* and *punctulatus* *Neopedies brunneri*	23 (XO)	—	22	SAEZ 1956
	Scottusa impudica and *lemniscata*	23 (XO)	—	22	SAEZ and SOLARI 1959
	Scottusa sp. A	21 (XO)	2	18	MESA 1960
	Aidemona aztecta	21 (XO)	2	18	SAEZ 1932
	Aleuas viticollis	19 (XO)	4	14	
	Aleuas brachypterus	19 (XO)	4	14	SAEZ 1956
	Philocleon anomalous	12 (neo XY)	10	—	HELWIG 1941
(ii) *Romaleinae*	Numerous species	23 (XO)	—	22	WHITE 1951, SAEZ 1956
	Diponthus maculiferus and *Ommaexecha servillei*	21 (XO)	2	18	SAEZ 1956
(iii) *Truxalinae*	*Pedioscirtetes maculipennis*	23 (XO)	—	22	WHITE and NICKERSON 1951
	Neopodismopsis abdominalis	19 (XO)	4	14	ROTHFELS 1950
	Many species	17 (XO)	6	10	WHITE 1951
Coleoptera	*Pissodes engelmanni, sitchensis* and *strobi*	34 (XY)	—	32	MANNA and SMITH 1959
	Pissodes affinis, dubius, fasciatus, notatus and *radiatae*	30 (XY)	4	24	
	Pissodes terminalis and *yosemite*	28 (XY)	6	20	
Diptera	*Drosophila virilis*	12 (XY)	—	10	PATTERSON and STONE 1952
	Drosophila texana	10 (XY)	2	6	
Reptilia					
(a) *Lacertilia*					
(i) *Agamidae*	*Uromastix hardwicki*	24	—	24	MATTHEY 1931, 1933
	Agama stellio	12	12	—	
(ii) *Anguidae*	*Gerrhonotus scincicauda*	22	2	20	
	Pseudopus apus	20	4	16	
(iii) *Geckonidae*	*Tarentola mauretanica*	42	—	42	
	Gekko japonica	38	4	34	
(iv) *Lacertidae*	*Lacerta*, most species	38	—	38	
	Lacerta ocellata	36	2	34	
(v) *Scincidae*	*Scincus officinalis*	32	4	28	
	Chalcides tridactylus	28	8	20	

Table 44 (Continued)

Group	Species	2n	Autosomes		Reference
			V's	Rods	
(b) *Crocodylia*	*Crocodylus porosus*	34	24	10	Cohen and Clark 1967
	C. niloticus	32	26	6	
Mammalia					
(i) *Ungulata*	*Goat*	60 (XY)	—	58	
	Domestic sheep	54 (XY)	6	46	
(ii) *Rodentia*	*Mus minutoides indutus*, *M. minutoides minutoides* ssp_1 and *Mus setulosus*	36 (XY)	—	34	Matthey 1964
	Mus minutoides musculoides	34 (XY)	2	30	
		32 (XY)	4	26	
	Mus minutoides minutoides ssp_4	32 (XY)	4	26	
Angiospermae	*Lycoris sanguinea*	22	—	22	Darlington 1963
	L. straminea	16	6	10	
	L. aurea	14	8	6	
		12	10	2	
	Fritillaria pudica	26	2	24	Darlington 1963
	F. imperialis and *meleagris*	24	4	20	
	F. latifolia, *nigra* and *ruthenica*	18	10	8	
Gymnospermae	*Podocarpus amarus*, *andinus*, *nivalis*, and *spicatus*	38	2	36	Hair and Beuzenberg 19
	P. ferrugineus and *ustus*	36	4	32	
	P. acutifolius, *hallii*, and *totara*	34	6	28	
	P. nagi	26	14	12	
	P. falcatus and *gracilior*	24	16	8	
	P. elongatus	22	18	4	
	P. henkelii	22	18	4	
		20	20	—	
	P. blumei, *comptonii*, *dacrydiodes*, *imbricatus*, *latifolius*, *minor*, *palustris*, *vieillardii*, and *vitiensis*	20	20	—	

Group	Species	2n	Autosomes		Reference
			V's	Rods	
	Dacrydium franklinii, intermedium, and *laxifolium* *Microcachrys tetragona*	30	10	20	HAIR and BEUZENBERG 1958
	Dacrydium biforme and *Saxegothaea conspicua*	24	8	16	
	Pherosphaera fitzgeraldi and *hookeriana*	26	14	12	
	Dacryidium kirkii	22	10	12	
	Dacryidium araucaroides, balansae, colensoi, cupressinum, elatum, guillaumindii, lycopodiodes and *taxoides* *Acinopyle pancheri*	20	20	—	

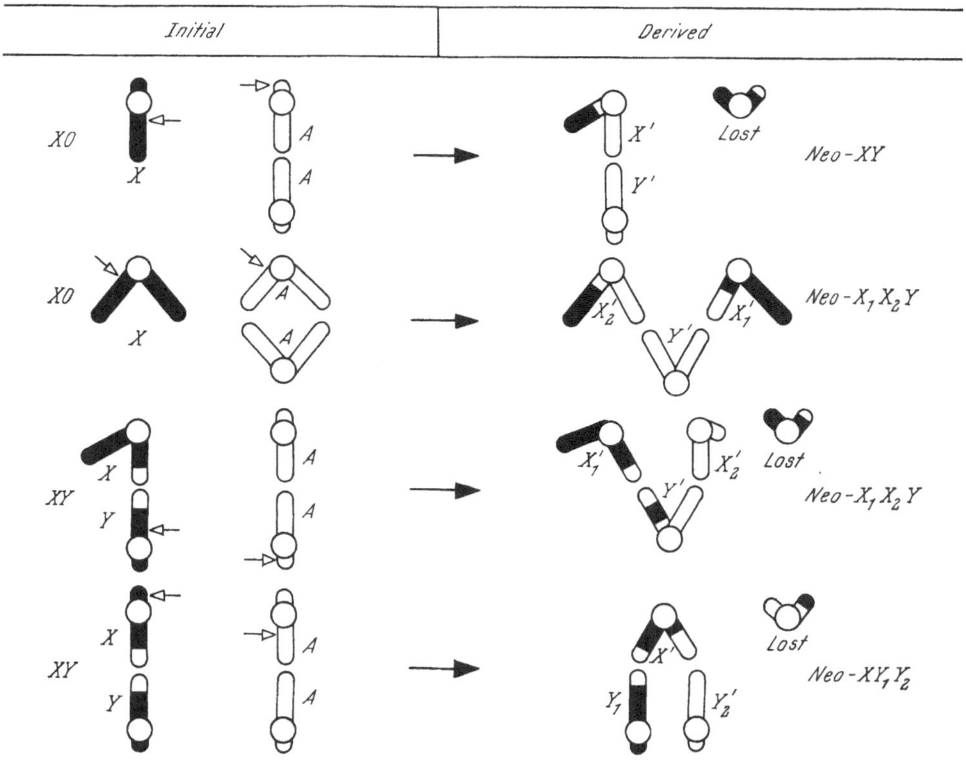

Fig. 60. The origin of neo sex-chromosome systems by centric fusion (after WHITE 1957).

In other plant cases too and especially in animals the relationships of chromosome complements between species within the same family are frequently suggestive of an equivalent process of reduction (Table 44). It must be admitted, however, that this interpretation involves a measure of subjectivity since it is just as easy to explain in terms of gain rather than loss and by a mechanism we have next to consider. The only certain cases of aneuploid reduction in animals are those involving the evolution of neo-sex chromosome systems (Fig. 60).

b) Mechanisms Leading to Increase

i) Misdivision (Centric Fission)

The compound nature of the localised kinetochore and its essentially duplicate nature (Lima-de-Faria 1956) means that metacentric elements can be divided by transverse breakage into two telocentric subunits. Such misdivision may follow failure of pairing (*Tulipa, Pisum, Nicandra*) or failure of co-orientation of paired chromosomes (*Fritillaria, Gasteria*). Provided that these telocentrics are stable, this will result in an increased chromosome number by producing two rods at the expense of one V. To this limited extent the behaviour of chromosomes with localised kinetochores can parallel that of those with non-localised centric systems (see pg. 126). Such a process of fission has certainly been seen under experimental conditions in *Campanula* (Darlington and La Cour 1950). And there is little doubt that the same process has led to the $2n = 14$ form of *Alisma plantago* (Fig. 61) and the $2n = 15$ *Spirea filipendula* (Maude 1940) which is regularly hybrid in respect of one V and 2 rods. Similarly *Tradescantia commelinoides* was first described by Celarier (1955) as $2n = 14$ metacentrics. Mattsson (1963), however, found $2n = 16$ with 12 metacentrics and 4 telocentrics. And, although telocentrics appear to be rare in plants, *Tradescantia micrantha* has a complement consisting solely of telocentrics (Fig. 62). Similarly, whereas most species of *Fritillaria* have 12 pairs of chromosomes consisting of 4 V's and 20 rods there are 13 chromosome pairs with only 2 V's in *F. pudica* (Darlington 1932).

It is possible, therefore, that some, at least, of the cases listed in Table 44 might be explicable in terms of centric fission. Part of the difficulty in resolving the situation stems from the fact that many rod-shaped chromosomes have yet to be accurately classified as to type. The mouse provides an instructive case. The mouse complement of 40 chromosomes has been variously described:

(1) All acrocentric (Chu 1961, Ohno and Lyon 1965).

(2) All telocentric (Levan, Hsu, and Stich 1962).

(3) All telocentric or nearly telocentric (Chu 1963).

Again, until recently, it has been the practice to regard the rod chromosomes of acrodoids as acrocentric (White 1954, 1957). There is now unambiguous evidence that some of them are definitely telocentric (John and Hewitt 1966 and see Fig. 42).

Be this as it may the many cases of supernumerary systems involving either telocentrics or else their iso-chromosome derivatives (Table 45)

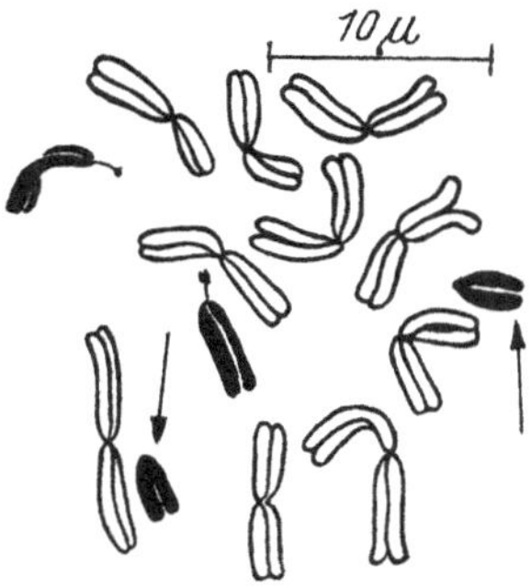

Fig. 61.

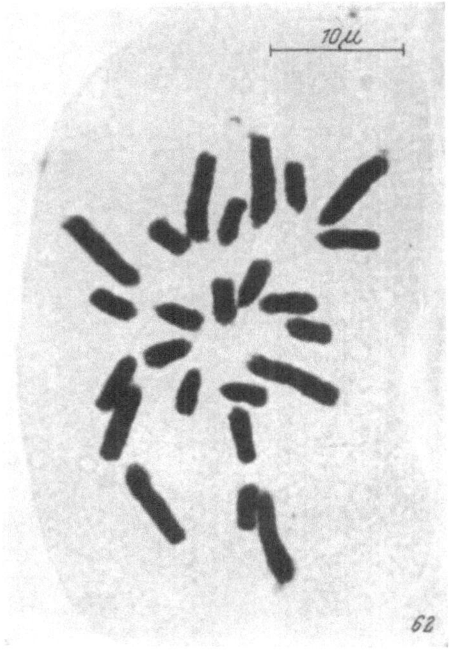

Fig. 62.

Fig. 61. The mitotic complement of a form of *Alisma plantago-aquatica* with $2n = 14$ showing two telocentric elements (arrows). $2n = 12$ forms are also known with 6 metacentric pairs (after CLAVIER unpublished as shown in DARLINGTON 1963).

Fig. 62. The mitotic complement of *Tradescantia micrantha* ($2n = 24$). All the members are telocentric. (Photograph kindly supplied by Dr. KEITH JONES).

Table 45. *Telocentric and Iso B-chromosomes in Plants.* (Compiled from BATTAGLIA 1964.)

Telocentric	Isochromosome	Telocentric and Isochromosome
1. *Allium pulchellum*	*Anthoxanthum aristatum*	*Allium cernuum*
2. *Anthurium* spp.	*Centaurea scabiosa*	*Lilium callosum*
3. *Caltha palustris*	*Chrysanthemum corymbosum*	*Lilium medeoloides*
4. *Paris tetraphylla*	*Chrysanthemum millefoliatum*	*Scilla scilloides*
5. *Poa alpina*	*Festuca pratensis*	*Secale cereale*
6. *Tradescantia virginiana*	*Poa trivialis*	*Sorghum purpureo-sericeum*
7. *Trillium* spp.		
8. *Xanthisma texanum*		

emphasise the effectiveness of misdivision as a potential mechanism for increasing chromosome number. It is true that such cases have not been

Table 46. *Polyploid Species in Angiosperr*

Genus	Species				
	2x	3x	4x	5x	6x
1. *Paris* (x = 5)	*hexaphylla* *obovata* *polyphylla* *tetraphylla*	*hexaphylla*	*quadrifolia*		
2. *Althaea* (x = 7)			*cannabina*		*cysterocarpa* *ficifolia* *officinalis* *pallida* *pontica* *sulphurea*
3. *Bromus* (x = 7)	*ciliatus* *laevipes*		*adoensis* *albidus* *arenarius* *grossus* *hordaceous* *interruptus* *lepidus* *mollis* *racemosus* *rubens* *texensis*		*auleticus* *brevis* *cappadocicus* *macrantherus* *sitchensis* *trinii* *uruguayensis* *variegatus*
4. *Festuca* (x = 7)	*amethystina* *borderi* *diviuscula* *geniculata* *paniculata* *polesica* *pumila* *rigida* *supine* *tatrae* *tenuifolia*		*altaica* *ciliata* *elmeri* *idahoensis* *occidentalis* *parvigluma* *subuliflora* *valesiaca* *vindula*		*arizonica* *coelestis* *cryophila* *fallax* *gigantea* *kirilovii* *longifolia* *sulcata*
5. *Potentilla* (x = 7)	*arguta* *glandulosa* *neptaphylla*		*alba* *erecta* *pulchella* *reptans*		*canescens* *nepalensis*
6. *Aconitum* (x = 8)	*excelsum* *lycoctonum* *orientale* *septentrionale* *uncinatum*	*stoerkianum*	*altaicum* *anglicum* *anthora* *delavayi* *firnum* *fischeii* *napellus* *volubile*		*palmatum*

(Based on DARLINGTON and WYLIE 1955.)

Species						
7x	8x	10x	12x	14x	16x	22x
	japonica					
		sinensis	*armenaica*			
	breviaristatus *carinatus* *gussonii* *maritimus* *pitensis*	*riparius*	*arizonicus*			
	californica *kingii*	*kryloviana* *maritima*				
	bifurca *curdica* *monspeliensis*	*norvegicus*		*sibthorpiana*	*haematochroa*	
	wilsonii					

Table 46 (Continued)

Genus	Species				
	2x	3x	4x	5x	6x
7. *Caltha* (x = 8)	*cornuta*		*palustris*		*leptocephala* *palustris*
8. *Delphinium* (x = 8)	*ajacis* *brachycentrum* *brunonianum* *cardinale* *decora* *nudicaule* *speciosum* *tatiense* *zalil*	*moerheimii*	*bulleyanum* *oxysepalum*		*belladonna* *lamartini*
9. *Draba* (x = 8)	*fladnizensis*		*incana*	*crassifolia*	*rupestris*
10. *Chrysanthemum* (x = 9)	*argenteum* *boreale* *cassium* *catanache* *cinerariaefolium* *mawii* *rotundifolium* *serotinum* *ulginosum*	*frutescens*	*corymbosum* *ircutianum* *oreades* *praealtum* *wakasaense*		*camphoratum* *ceratophylloid.* *japonense* *shimotomaii* *sibiricum* *silvaticum* *sinense* *weyrichii*
11. *Dioscorea* (x = 10)	*caucasia* *gracillima* *quinquelobe* *tokoru*		*discolor* *macroura* *japonica*		*sativa* *villosa*
12. *Optunia* (x = 11)	*brasilaris* *brasiliensis* *chlorotica* *microdasys* *repens* *santa-rita*	*monacantha*	*leucotricha* *salmiana* *tomentosa*		*dillenii* *fragilis* *subulata*
13. *Danthonia* (x = 12)	*alpicola* *auriculata* *carphoides* *duttoniana* *monticola* *occidentalis* *semiannularis* *setacea*		*clelandii* *erjantha* *geniculata* *richardsonii*		*bipartita* *induta* *laevis* *pallida* *purpurascens*

Species						
7x	8x	10x	12x	14x	16x	22x
palustris	*polypetala*					
	aurea *dahurica* *sachalinensis*	*arctica* *alpina*	*arabisans*	*alpina*		
rubellum	*arcticum* *decaisneanum* *ornatum* *zawadskii*	*maximum* *pacificum* *yezoense*		*maximum*		*lacustre*
	bulbifera			*cayennensis* *oppositifolia*		
	ficus-indica	*cylindrica*				
	induta *pallida* *procera*	*induta*				

Table 46 (Continued)

Genus	Species				
	2x	3x	4x	5x	6x
14. *Deutzia* (x = 13)	*gracilis* *hypoglauca* *sieboldiana*				*mollis*
15. *Rhododendron* (x = 13)	305 species	*diaprepes* *russatum*	30 species		*artosquamatum* *cinnabarium* *complexum* *concatenans* *cinnabarium* *cuneatum* *davidsonianum* *heliolepis* *keysii* *manipurense* *polyandrum* *siderophyllum* *tapetiforme* *timeteum* *xanthocodon* *yunnanense*
16. *Nymphaea* (x = 14)	*capensis* *stellata*		*flava* *lotus* *rubra*		*odorata* *tuberosa*
17. *Magnolia* (x = 19)	21 species	*stellata* *rubra*	*accuminata* *cordata* *liliiflora*	*soulangerna*	*campbellii* *dawsoniana* *denudata* *grandiflora* *mollicomata* *sargentiana* *schiedeana* *sprengeri*

directly stabilised in nature. In *Nicandra physaloides,* however, a pair of iso-chromosomes are found together with nine ordinary pairs. One of these iso's is subject to loss at meiosis but the other forms an indispensable and permanent member of the set. Plants with 20 and 19 chromosomes are found but seed with 20 chromosomes germinates earlier than that with only 19 (DARLINGTON and JANAKI-AMMAL 1945). Likewise in *Philadelphus* the long metacentrics in the hybrid species *P. lemoinei* and *P. zeyheri* frequently fail to pair and the univalents produced form rings showing them to be iso-chromosomes (JANAKI-AMMAL 1958). Evidently then stable iso-chromosomes can be developed. HAGA (1961) too comments on the fact that although a minute telocentric B-chromosome varied among cells its iso-chromosome derivative was perfectly stable.

Species						
7x	8x	10x	12x	14x	16x	22x
	discolor *reflexa* *vilmorinae*	*crenata* *schneideriana*				
	pholidotum		*manipurense*			
	tetragona				*gigantea*	

WHITE (1957) is of the opinion that misdivision is unlikely to have played an important role in evolutionary increases of chromosome number. Unquestionably this stems from his belief that the centromere in animal chromosomes "always seems to occupy an interstitial position and never a terminal one." In order to achieve such an increase he has therefore proposed a mechanism of dissociation whereby the centromere and the telomeres of "donor" chromosomes are utilised to foster the formation of two separate linkage units from the two arms of a metacentric (see Fig. 39). And B-chromosomes, of course, immediately suggest themselves as suitable donors. If this process really does occur—and the direct evidence for it is far from satisfactory—then it offers an indirect means of stabilising supernumerary material.

ii) Polyploidy

Apart from indirect aneuploid increase, the entire chromosome complement can be multiplied geometrically—either directly or more usually following hybridisation—resulting in states of polyploidy. Polyploids are most numerous in angiosperms (Table 46) though they are found throughout the plant kingdom. Assuming n = 7–9 to be basic, then numbers in

Table 47. *Polyploid Species in the Oligochaeta.*
(Data of Omodeo 1951, 1952; Muldal 1952 and Christensen 1961.)

Family	Species	Level of Ploidy							
		2x	3x	4x	5x	6x	7x	8x	10x
Family *Lumbricidae*	*Bimatus eiseni*	32							
	tenuis		48						
	Dendrobaena mammalis	34							
	rubida			68					
	subrubicunda			68					
	rubida tenuis					102			
	octaedra	34		68			ca. 126		
	Eisenia veneta	36							
	rosea typica		54						
	rosea dendrabaenoides			72					ca. 180
	Octolasium lacteum	38		76					
	croaticum typicum							140	
	cyaneum								ca. 190
Family *Enchytraeidae*	*Enchytraeus lacteus*	18							
	bucholzi	18		36	54	72			
	Mesenchytraeus armatus, beaumeri, flavus and *glandulosus*	16							
	pelicensis			32					

excess of 14 are in some degree polyploid and, on this basis, Grant (1963) has estimated that some 43% of nearly 12,000 dicotyledonous species and 58% of over 5,000 species of monocotyledonous which have been studied are polyploid. This gives an average percentage polyploidy of 47%. Nearly 75% of the *Gramineae* are polyploid and a high frequency of polyploids is found also among the *Rosaceae, Polygonaceae, Malvaceae, Crassulaceae, Nymphaceae,* and *Arabaceae.*

As we saw earlier (see pg. 64) several cases of intra-specific polyploidy are on record in enchytraeid and lumbricid worms. Interspecific polyploid relationships are also known in these two families (Table 47 and Fig. 63). In addition a case exists for arguing that polyploidy may have occurred in earwigs, beetles and in the orthopteran genus *Gryllotalpa* (Table 48). The case is strongest in earwigs where the chromosome numbers fall into a lower series with $2n = 12$–14 and a higher one with $2n = 24$–25. In the latter case four of the five species possess X_1X_2Y multiple sex chromosome systems in some or all of their members suggesting that they may be tetra-

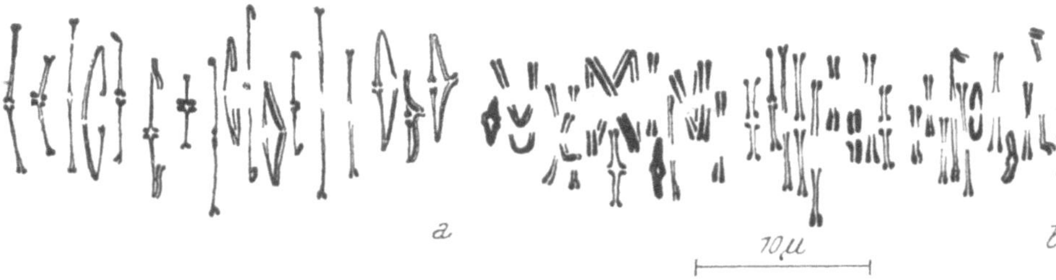

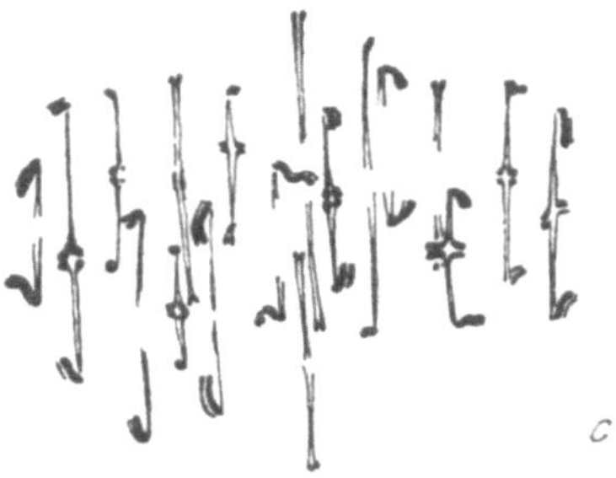

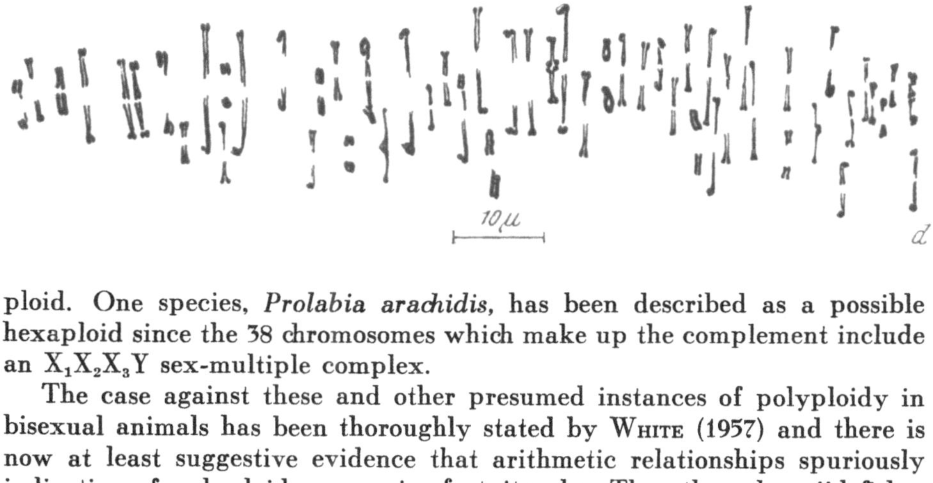

Fig. 63. Polyploid species in the earthworm genera *Dendrobaena* Figs. 63*a* and *b*) and *Eisenia* (Figs. 63*c* and *d*). All four cells are first meiotic metaphases from eggs (after MULDAL 1952).
a — *D. mammalis* ($2n = 34$)
b — *D. subrubicunda* ($2n = 68$)
c — *E. venata* ($2n = 36$)
d — *E. rosea* ($2n = 54$)

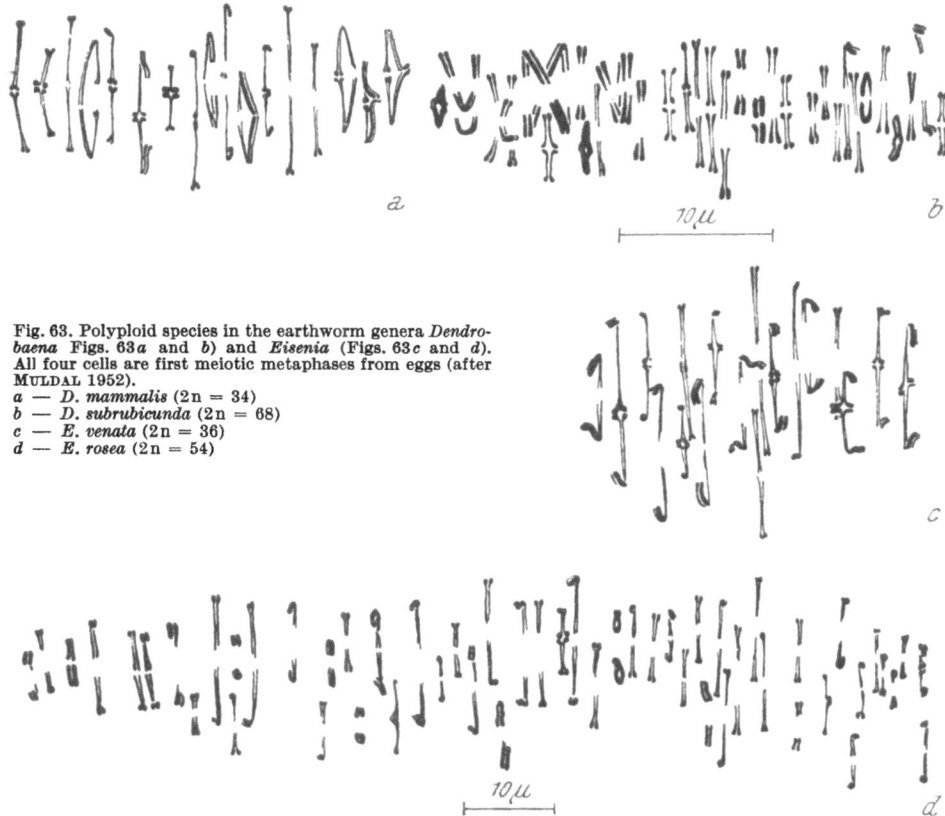

ploid. One species, *Prolabia arachidis,* has been described as a possible hexaploid since the 38 chromosomes which make up the complement include an $X_1X_2X_3Y$ sex-multiple complex.

The case against these and other presumed instances of polyploidy in bisexual animals has been thoroughly stated by WHITE (1957) and there is now at least suggestive evidence that arithmetic relationships spuriously indicative of polyploidy can arise fortuitously. Thus the salmoniid fishes *Salmo salar* ($2n = 60$) and *S. trutta* ($2n = 80$) which SVÄRDSON (1945) suggested to be hexaploid and octoploid respectively turn out to have a similar mean DNA content (REES 1964). Likewise HUGHES-SCHRADER (1958) has shown how a Robertsonian equivalence of two karyotypes, subsequently masked by pericentric inversions, can simulate a polyploid relationship. On the other hand very strong evidence in support of regular tetraploidy

Table 48. *Species Relationships Suggestive of Polyploidy in Bisexual Animal Species.* (Based on Goldschmidt 1953 and Ortiz 1968).

Group	Species	2n (♂)	Sex Chromosome System
	Labidura bidens	12	XY
	riparia	14	XY
	Labia minor	14	XY
	Anechura bipunctata	24	XY
	Apterygida albipennis	24	XY
	Forficula scudderi	24	XY
Forficulina	*smyrnensis*	21	X_1X_2Y
(Dermaptera)	*auricularia*	24	XY
		25	X_1X_2Y
	Anisolabis annulipes	25	X_1X_2Y
	marginalis	25	X_1X_2Y
	maritima	25	X_1X_2Y
	Euborellia moesta	25	X_1X_2Y
	Pseudochelidura sinuata	25	X_1X_2Y
	Nala lividipes	37	X_1X_2Y
	Prolabia arachidis	38	$X_1X_2X_3Y$
	Gryllotalpa gryllotalpa	12 (W. Europe)	XY
		14 (Roumania)	XY
Orthoptera		15 (S. Italy)	XO
		19, 23	XO
	africana	23	XO
	borealis	23	X_1X_2Y
	Blaps lusitanica	19	X_1X_2Y
	lethifera	37	X_1X_2Y
	mortisaga	36	$X_1X_2X_3Y$
Coleoptera	*mucronata*	36	$X_1X_2X_3Y$
	gigas	35	$X_1X_2X_3X_4Y$
	cribosa	36	$X_{1-12}Y_{1-6}$
	polychestra	36	$X_{1-12}Y_{1-6}$

has recently been presented in the bisexual south american frog *Odontophrynus americanus* (Beçak, Beçak and Rabello 1966). This evidence is of 3 kinds:

(a) The 44 chromosomes present in the complement of this species fall into eleven groups consisting of four homologues per group.

(b) At meiosis eleven quadrivalents, or else their equivalent breakdown products, were regularly observed, and

(c) Related species from Argentina (*O. occidentalis*) and Brazil (*O. cultripes*) have exactly half the chromosome number ($2\,n = 22$).

In a further species of the same family, *Ceratophrys dorsata*, Beçak, Beçak and Rabello (1967) record a count of $2\,n = 104$ in which the karyotype consists of 13 groups of 8 homologues each. At meiosis mixtures of octavalents, hexavalents, quadrivalents and bivalents are found suggesting that this species is an octoploid. Significantly $2n = 26$ diploids are known in the related species *Chacophrys pierotti* and *Lepidobatrachus ilanensis.*

Evidence in favour of polyploidy in fishes has been given by SUOMALAINEN (1958) but its value remains debatable especially in view of the complex mosaicism reported for some fish species (see pg. 43). Extensive polyploidisation in fish has also been suggested by OHNO and ATKIN (1966) to account for the large DNA differences which exist between them. Indeed they have made the remarkable generalisation that "in Vertebrates, any DNA values above 20% that of placental mammals indicate polyploid lineages." OHNO (1968) has also argued that a diploid/tetraploid relationship exists between the clupeoid and salmonoid fishes of the pacific. In support of this is the difference in chromosome number, DNA content and the fact that while the presumed diploid has two separate gene loci which code for the A and B sub-units of lactate dehydrogenase both loci are present in duplicate in the presumptive tetraploid. In view of the evidence to be discussed in the next section the value of the DNA data, as it stands, is, to say the least, a matter for considerably more objective appraisal than it has so far received.

2. Differences in Structure

As we have seen structural changes are clearly and directly implicated in many of the cases where numerical differences exist between related species. They may also be involved in cases where no such differences

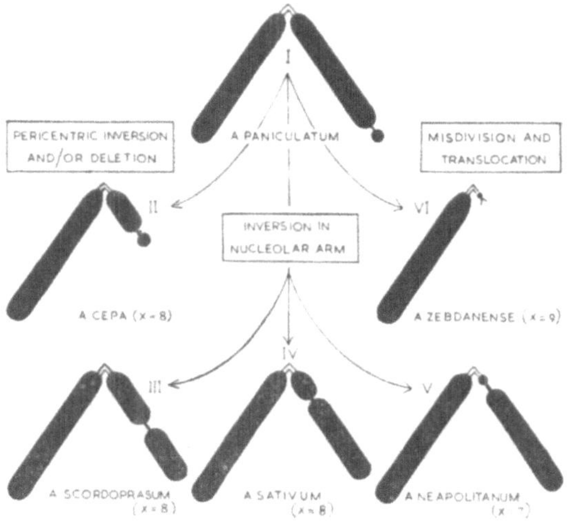

Fig. 64. The six main types of nucleolar chromosome found in the genus *Allium* and their presumed relationships as inferred by VED BRAT (1965).

occur or else where they can have played no direct role in the establishment of them. In the genus *Allium* for instance the nucleolar chromosomes vary in different species and six main types can be distinguished (VED BRAT 1965). These different types and their suggested inter-relationships are summarised in Fig. 64. This scheme is based on the assumption that the metacentric type with a satellite, which is present in some 33% of the species with x = 7 and 8, constitutes the basic type.

An alternative indication of structural change can sometimes be provided by the study of nuclear DNA content and chromosome size. We have already seen how chromosome size may vary, often considerably, within the individual according to cell type (see pg. 24). For this reason similar cells must be used in inter-specific comparisons of this sort. Taking this and other necessary precautions Rees and Jones (1967) find that a DNA difference between two species of *Lolium* with the same chromosome number and

Table 49. *The Relationships Between Total Chromosome Volume, Nuclear DNA Content and Nuclear Dry Mass in the Genus Lolium.*
The measurements were made on 2C nuclei of root apices.
(Data of Rees and Jones 1967.)

Type	2n	Volume	DNA	Mass	DNA/Volume	DNA/Mass	Mass/Volume
1. *Lolium perenne*	14	25.9	31.3	3.40	1.28	9.2	0.13
2. *L. temulentum*	14	37.1	42.1	4.67	1.13	9.0	0.13
3. F_1 hybrid	14	30.7	36.8	4.19	1.20	8.8	0.14

Table 50. *DNA Relationships in the Genus Vicia.*
Each value represents the mean of 50 readings.
(Data of Martin and Shanks 1966.)

Species	2n	Relative DNA Content
1. *Vicia faba*	12	100 (standard)
2. *Vicia hybrida*	12	50.93 ± 1.42
lutea	14	59.006 ± 1.71
narbonensis	14	56.75 ± 1.34
3. *Vicia amphicarpa*	10	16.74 ± 0.37
sativa	12	18.54 ± 0.51

similar karyotypes is paralleled by differences both in chromosome volume and nuclear dry mass (Table 49). And pachytene pairing relationships in F_1 hybrids clearly implicate simple structural differences of the duplication/deficiency type in this case (see for example Fig. 131 in John and Lewis 1965).

Differences in DNA content are found also in different diploid species of the genus *Lathyrus* which share 14 chromosomes but which vary considerably in chromosome size and mass (Rees *et al.* 1966). The genus *Vicia* too (Martin and Shanks 1966) shows startling differences in DNA level between its member species (Table 50). The structural basis for these differences is not clear. Perhaps they parallel the distinctive process of duplication which Keyl has found to operate between subspecies of *Chironomus thummi* (see pg. 91). For example Ullerich (1966) has com-

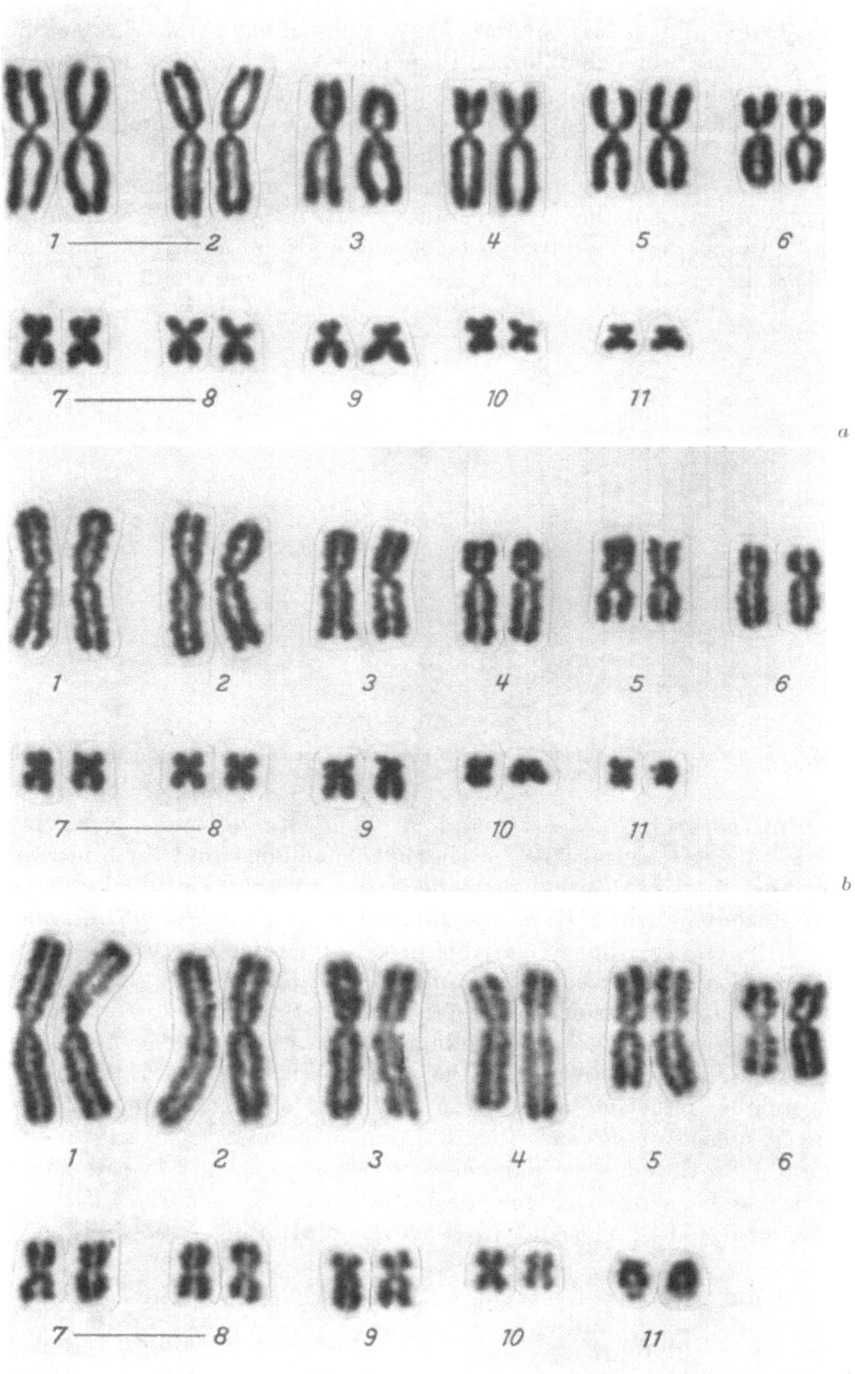

Fig. 65. The diploid karyotypes ($2n = 22$) of *Bufo bufo* (Fig. 65*a*), *B. viridis* (Fig. 65*b*) and *B. calamita* (Fig. 65*c*) as seen at spermatogonial metaphase (after ULLERICH 1966).

pared the DNA levels of 3 species of *Bufo* all with $2n = 22$ (Fig. 65). These three toads all possess similar karyotypes though the chromosomes of *Bufo bufo* are somewhat larger than those of *B. viridis* and *B. calamita.* All chromosomes of *B. bufo* were found to contain significantly more DNA than their homologues in the two other species (Fig. 66) and, of particular significance, the same DNA-differences between both sets of chromosomes were found also in hybrids between *B. bufo* and *B. viridis.* Significant differences between the DNA content of *B. viridis* and *B. calamita* were found only between the large chromosomes and the ratio of the total amount of DNA in the 3 genomes was 1.49 : 1.07 : 1.00. The comparable ratio for

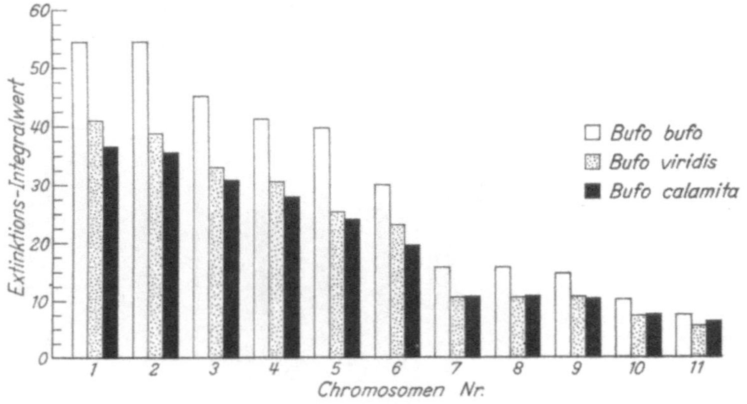

Fig. 66. The relative DNA content of the eleven members of the chromosome complements of *B. bufo, viridis* and *calamita* respectively (after Ullerich 1966).

Rana temporaria, R. arvalis and *R. esculenta,* all with $2n = 24$, was 1.00 : 1.28 : 1.54 (Ullerich 1967). Ullerich concluded that these interspecific differences in DNA content depended upon just the kind of local increase in DNA content which Keyl had found between *Ch. thummi thummi* and *thummi piger.* To what extent this process operates in other cases remains to be clarified. But even if this and the more typical duplication process found in *Lolium* are ruled out there remains the possibility of differential polynemy as a means of producing differences in DNA-content. Such a process has been implicated in those cases where a 2 : 1 ratio in chromosome number is accompanied by a 1 : 1 ratio of DNA values as has been found in species of *Thyanta* (Schrader and Hughes-Schrader 1958). Finally 1 : 1 ratios in chromosome number with accompanying 2 : 1 ratios in DNA, such as has been reported for species of *Bucholzia* and *Enchytraeus,* have been interpreted by Christensen (1966) to imply differential polyteny.

3. Variation in Systems with Non-Localised Kinetochores

Where the kinetochore is a localised entity the possibilities for chromosome evolution are limited by the fact that only chromosome fragments with centromeres—or, in some cases (see pg. 112), fragmented centromeres—can survive. In systems where the kinetochore is non-localised, on the other hand, it has been shown on a number of occasions, and in both plant and animal species, that fragments resulting from irradiation survive through successive mitoses. For this reason it is commonly believed that in such systems simple fragmentation can lead to numerical changes of evolutionary potential and hence to intergenomic differentiation.

In support of this belief is the range of chromosome numbers and DNA contents to be found in the Juncaceae. Thus in the *campestris—multiflora* complex of the genus *Luzula* the basic chromosome number is $2\,n = 2\,x = 12$

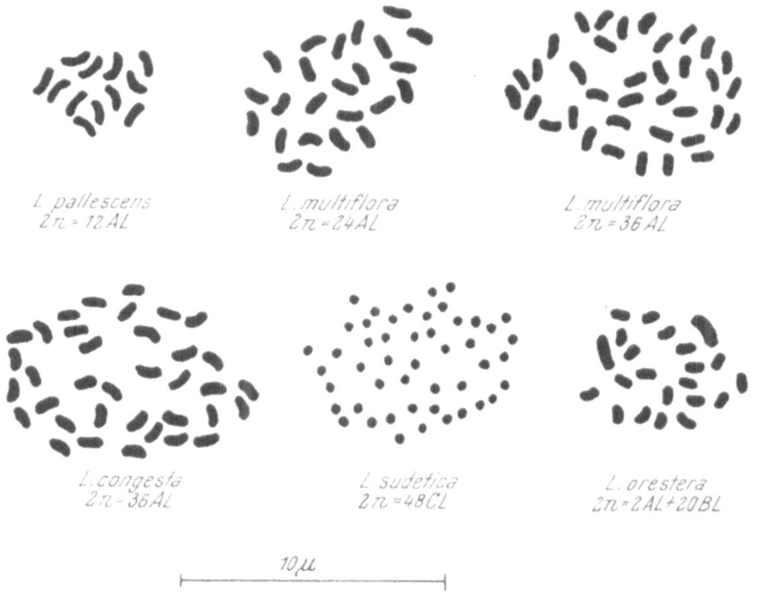

Fig. 67. Interspecific variation of the chromosome complement in the genus *Luzula* (after NORDENSKIÖLD 1961).

as found in *Luzula pallescens* and *L. campestris.* Some species show true (numerical plus quantitative) polyploidy. Thus *L. multiflora* is tetraploid ($2\,n = 4\,x = 24$) or hexaploid ($2\,n = 6\,x = 36$) and *L. congesta* has the octoploid number ($2\,n = 8\,x = 48$). In all these cases the chromosomes both within and between these complements are approximately equal in size and are referred to as standard or AL-types (Fig. 67). Other species of the complex while also showing polyploid numbers have chromosomes of a different size. Thus *L. australica* has 24 half-sized or BL-chromosomes while *L. sudetica* has 48 quarter-sized or CL-chromosomes. Because the total chromosome length in these complements appears to be the same as in the diploids, it is assumed that they are not true polyploids but rather endonuclear polyploids produced by transverse fragmentation of the basic 12 AL diploid. This is supported by the fact that while they cross readily with the diploids they do not cross with true polyploids which have the same number of chromosomes as themselves. It is also supported by DNA

Table 51. *DNA Relationships in the Genus Luzula.*
(Data of Mello-Sampayo 1961 and Halkka 1964.)

Species	2n	Chromosome Types	DNA Value	
			2C	4C
1. *Luzula luzuloides*	2n = 2 x = 12	all AL	3.22	6.85
2. *Luzula pallescens*	2n = 2 x = 12	all AL	2.80	4.62
3. *Luzula multiflora*	2n = 4 x = 24	all AL	5.03	9.12
	2n = 6 x = 36	all AL	6.32	12.09
4. *Luzula congesta*	2n = 6 x = 36	all AL	6.45	12.40
5. *Luzula sudetica*	2n = 8 x = 48	all CL	2.94	4.50
6. *Luzula orestera*	2n = 2 x + 10 = 22 (decasomic)	2AL + 20BL	3.70	6.75
7. *Luzula purpurea*[1]	2n = 2 x = 6	6 very large chromosomes	9.55	19.15

[1] N. B. The complement of *L. purpurea* is not easily reconciled with the other species in the genus. It has the lowest chromosome number but the largest chromosomes and the highest DNA content.

relationships within the complex (Table 51). It would appear, therefore, that two lines of chromosome evolution have occurred in the genus *Luzula:*

(a) Numerical plus quantitative polyploidy
12 AL (2 x) → 24 AL (4 x) → 36 AL (6 x) → 48 AL (8 x).

(b) Numerical/structural polyploidy
12 AL (2 x) → 24 BL (4 x) → 48 CL (8 x).

A further finding in support of this concept of transverse fragmentation is the occurrence, in other species of the complex, of aneuploid chromosome numbers in which some of the standard AL-members are replaced by two half-sized BL or four quarter sized CL-chromosomes. For example in *L. campestris* endo-tetrasomic and endo-hexasomic types are known with 2 n = 14 (10 AL + 4 BL) and 2 n = 16 (8 AL + 8 BL) respectively. In *L. orrestera* endo-octosomic and endo-decasomic types occur which have 2 n = 20 (4 AL + 16 BL) and 2 n = 22 (2 AL + 20 BL). Like the endopolyploids these aneuploids cross with the diploids but not with the true polyploids of the complex.

From a study of experimental hybrids between diploid, endopolyploid and aneuploid types Nordenskiöld (1961) has been able to support the conclusions reached on morphological grounds that BL and CL chromosome types correspond to, and are homologous with, parts of the AL chromosomes of diploids. In such hybrids, as indeed in natural strains, 2 BL or 4 CL chromosomes replace and correspond with 1 AL. The net result is the production at meiosis of 1 AL–2 BL and 1 AL–4 CL associations.

Since 2 BL and 4 CL chromosomes have always been found to correspond to, and to pair with, one large AL of the basic diploid set, the most likely origin of these BL and CL types seems unquestionably to be that of transverse fragmentation of the AL members. What is more difficult to appreciate is why the fragmentation process should have divided the chromosomes so regularly as to produce exactly halved entities.

Carex, like *Luzula,* has a non-localised kinetochore system. It also has a long inter-specific aneuploid series (Table 52) while short aneuploid series also exist within *C. caryophylla* and *C. digitata.* DAVIES (1956) is of the opinion that the diffuse nature of the kinetochore, and hence the possibility of producing stable and viable fragments, accounts for this extensive aneuploidy. In support of this mode of origin is the fact that *C. elata* (37 chromosomes), *aquatilis* (38 chromosomes), *caespitosa* (40 chromosomes) and *nigra* (42 chromosomes) have very small chromosomes while in *pilulifera* (9), and *digitata* (24) the chromosomes are relatively large.

Table 52. *Chromosome Numbers in the Genus Carex.*
(Data of DAVIES 1956.)

Species	n	Species	n
Carex pilulifera	9	*C. ovalis* and *pallescens*	32
C. ericetorum	15	*C. caryophyllea*	32, 33, 34
C. panicea	16	*C. strigosa*	33
C. humilis	18, 36	*C. bigelowii, lepidocarpa,*	
C. montana	19	*mairii* and *punctata*	34
C. depauperata	22	*C. demissa, pediformis,*	
C. digitata	24, 25, 26	*scandinavica* and *serotina*	35
C. bicolor	25	*C. laevigata*	36
C. dioica	26	*C. binervis, distans*	
C. atrata	27	and *elata*	37
C. hostiana	28	*C. aquatilis* and *flaccas*	38
C. divulsa and *lachenalii*	29	*C. caespitosa*	40
C. extensa and *flava*	30	*C. nigra*	42
C. distacha, peregrina		*C. hirta*	56
and *remota*	31		

STRANDHEDE (1965) has uncovered a remarkable cytological instability in natural populations of species belonging to the genus *Eleocharis* subseries Palustres. Heteroploid plants are common. They are, however, heterogeneous and include trisomics, hybrid segregants, fragmentation types and fusions. In all these aberrant types viability and fertility were good. The point of particular importance with regard to the issue under discussion is that here also long chromosomes can undergo transverse fragmentation to produce two medium to small derivatives.

The same principle appears to apply to animal species as well. Thus among the *Lepidoptera* the lowest haploid number is 8 (*Erebia tyndarus*) and the highest is 191 (*Lysandra nivescens*) while the most common haploid numbers are 29—31. The few DNA measurements available (Table 53) show that the DNA content of closely related species is almost equal despite considerable differences in chromosome number. And this, taken in conjunction with the fact that the group is characterised by a non-localised kinetochore, suggests that numerical differences within the Lepidoptera must depend upon fusions and/or fragmentations.

White (1957) has argued that the process of transverse fragmentation as applied to the genus *Luzula*—and indeed elsewhere—neglects the telomere concept. The process of chromosome diminution that occurs regularly in *Parascaris* shows convincingly that telomeres are not equally important in all animals and suggests that the telomere concept may not apply in non-localised systems. Indeed there are indications that it cannot always be

Table 53. *DNA Relationships in Moths.*
Each value represents a mean of 3–5 first meiotic ♀ metaphases.
(Data of Suomalainen 1965.)

Species	n	DNA Content (4C)
1. *Cidaria obeliscata* (a)	13	6.70
junipera (b)	30	6.13
2. *Cidaria minna* (c)	17	5.94
suffumata (d)	32	7.45

Table 54. *Chromosome Numbers in the Armoured Scale Insects.*
The numbers in parentheses indicate the number of species in which spontaneous fragmentation has been observed.
(Data of Brown 1960.)

Tribe	No. of Species with Diploid Complements of						Total	Total No. of Species in which Spontaneous Fragmentation Seen
	8	10	12	14	16	18		
1. *Diaspidini*	14 (3)	2 (1)	2	—	2 (1)	3	23	5
2. *Aspidotini*	36 (12)	1	—	—	—	—	37	12
Totals	50 (15)	3 (1)	2	—	2 (1)	3	60	17

easily applied even in localised systems. Thus in both plants and animals, whether holo- or mono-kinetic, the fate of broken ends is variable and depends, in part at least, on the tissue in which the break occurs. For example in maize endosperm chromatids fuse *inter se* at the site of the break to give rise to a breakage-fusion-bridge cycle. In maize embryos, however, the broken surface forms a new and stable end. Similarly in the coccid *Icerya purchasi,* fusing and bridging follows radiation-induced breakage in spermatocytes but not in somatic or embryonic cells (Hughes-Schrader and Ris 1941).

Suomalainen and Hallka (1963) have pointed out that the presence of complex sex-determining loci in animals with non-localised kinetochores, in both sex-chromosomes and autosomes, could well interfere with transverse fragmentation. Indeed it is in these terms they explain the fact that in both *Lepidoptera* and *Hemiptera* the deviations from the presumed modal number are not as common or as extensive as that found in *Luzula.* In all probability the true situation is more subtle. Thus the armoured scales (*Diaspididae*) are characterised by a non-localised kinetochore and fragmentation of chromosomes by X-irradiation produces perfectly functional fragments. Despite this the range of numerical variation in the group is not especially high (Table 54). Moreover, although 8 is obviously the basic diploid number for both of the major tribes in the armoured scales yet the two differ dramatically in the range of chromosome number which they show. Indeed, apart from *Comstockiella* the aspidiotines show no deviation from the basic number of 8. Despite this Brown (1960) found that spontaneous fragmentation occurred at approximately the same extent in both major tribes (Table 54). If spontaneous fragmentation has provided the raw material for increase in chromosome number then that material has been about equally available in both major tribes but has not been equally utilised by them.

VI. The Induction of Genetic Changes in the Karyotype

1. Irradiation and Chemicals

Both physical agents, like ionising radiations, and chemical substances produce changes in chromosome structure and organisation. Though the mode of action of these agents may be quite different they produce apparently similar morphological changes. Thus if the unit of breakage or reunion is used as the basis for classification then three types of aberration are distinguishable:

(a) Chromosome aberrations.

(b) Chromatid aberrations.

(c) Half-chromatid aberrations.

The first of these is induced before replication of the chromosomes in interphase nuclei (G_1-phase). The second category is the result of induction during or after the phase of replication (S-phase) while the third is found only following the treatment of prophase.

The types of structural change that can occur within or between chromosomes and chromatids are of two principal kinds—the simple or terminal deletion and the exchange. Let us begin with chromosome-type aberrations (Fig. 68) since there are fewer exchange types to consider.

Exchanges may occur within a chromosome (intra-change) or between chromosomes (inter-change) and may be complete or incomplete, in which case two open ends remain ununited. A complete exchange within a chromosome arm results either in an ordinarily undetectable paracentric inversion or an intercalary deficiency. An incomplete exchange gives rise to what appears to be a simple deletion. Inter-arm intra-changes are of two types. One (U-type) results in centric rings together with fragments and these have been termed asymmetrical intra-changes. The other (X-type) gives rise either to a symmetrical configuration, which may be undetected if complete, and to a deletion if incomplete. Simple chromosome inter-changes are also classified into asymmetrical and symmetrical types, complete aberrations of the former category resulting in dicentrics.

Chromatid aberrations are similar in many respects to the chromosome types although there may be a greater variety of changes including inversions, duplications, deficiencies and interchanges (*sensu stricto*) (Fig. 69).

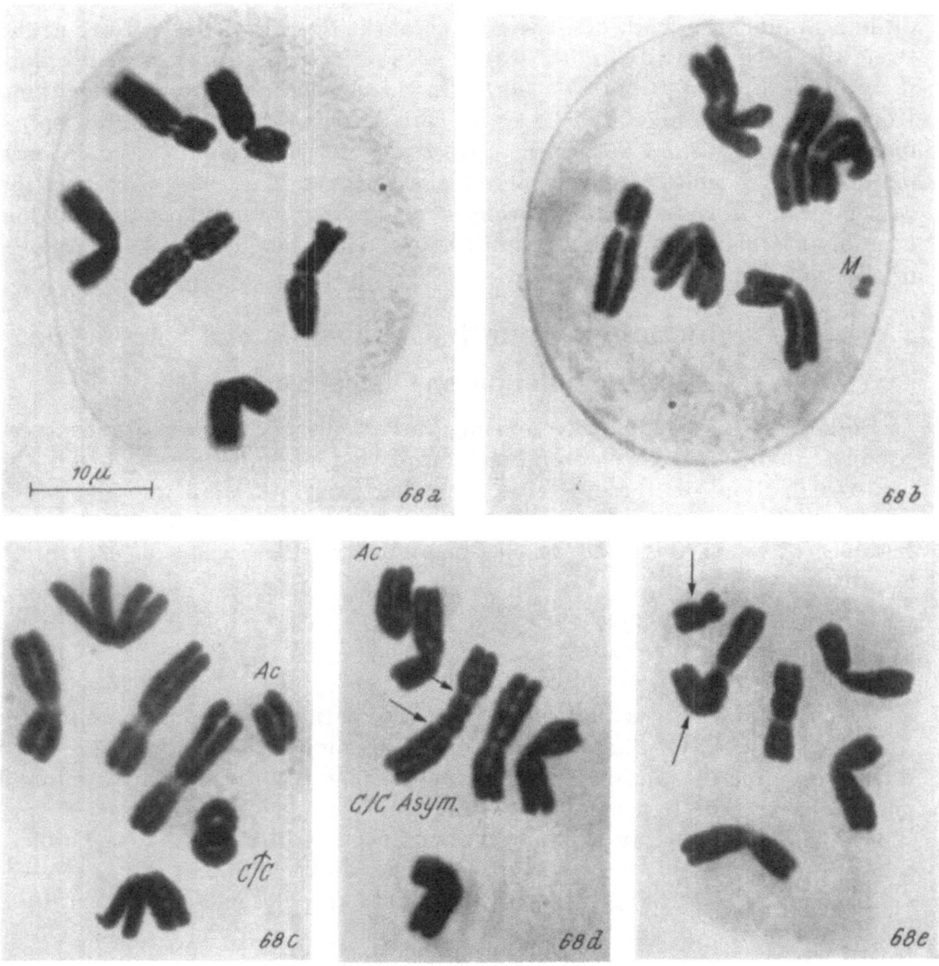

Fig. 68. X-ray induced chromosome-type aberrations in pollen grains of *Tradescantia bracteata* (photographs *c*, *d* and *e* kindly supplied by Dr. G. H. Jones).
a — Normal haploid complement (n = 6)
b — Paired minutes (M) arising from interstitial deletion
c — Paired centric rings (C͡C) and an acentric chromosome fragment (AC)
d — A dicentric (C/C Asym.) chromosome (arrows) and an accompanying acentric chromosome fragment (AC)
e — A reciprocal translocation (C/C Sym.) leading to the production of two new chromosome types (arrows).

Part of the reason for this greater variety is due to the increase in the possible types of intra-change at the chromatid level. Thus exchanges can occur between sister chromatids at similar loci—so-called iso-chromatid aberrations. More important, however, is the fact that, because of the pairing between sister chromatids, the symmetrical complete types of

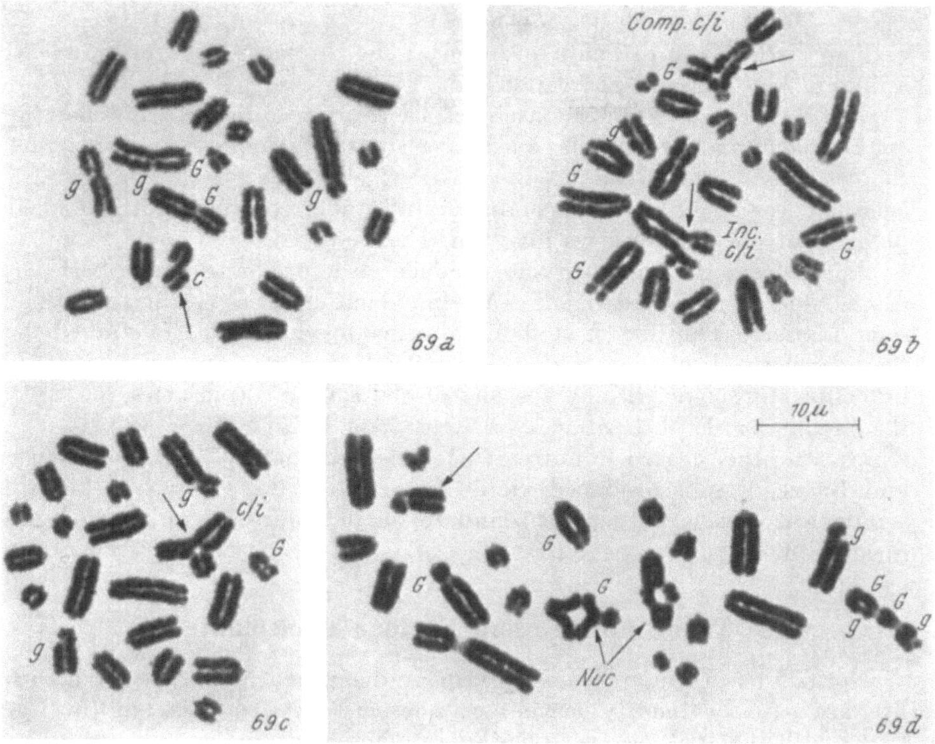

Fig. 69. X-ray induced chromatid-type aberrations in embryonic cells of the desert locust, *Schistocerca gregaria* (2n♀ = 22 + XX, ♂ = 22 + XO). Note numerous acromatic lesions, both single gaps (*g*) and paired iso-gaps (G) are present in all four cells.

a — A simple chromatid break (*c*) in a ♂ embryo
b — One complete (com.) and one incomplete (inc.) chromatid-iso-chromatid exchange (c/i) in a ♀ embryo
c — One complete chromatid-isochromatid exchange (com.c/i) in a ♂ embryo
d — Two NUC-type exchanges (see Fig. 70) together with one insertion intrachange (arrow).

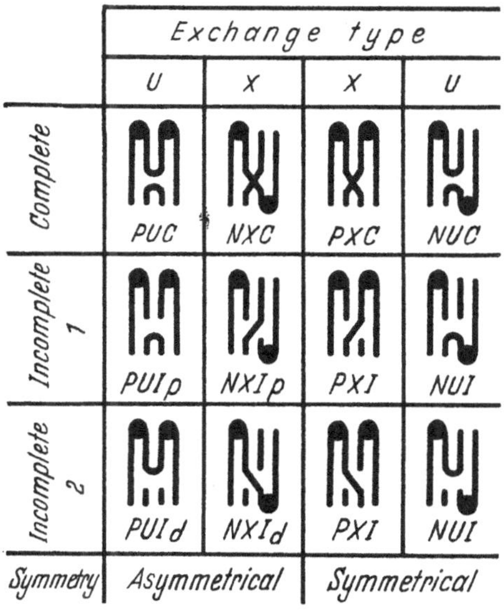
Exchange type
U
X
X
U
Complete
PUC
NXC
PXC
NUC
Incomplete 1
PUIp
NXIp
PXI
NUI
Incomplete 2
PUId
NXId
PXI
NUI
Symmetry
Asymmetrical
Symmetrical

Fig. 70. Classification of chromatid exchange types (after Fox 1967).

exchange which are difficult to detect at the chromosome level are clearly apparent in chromatid aberrations.

The inter-arm chromatid exchanges are similar to their chromosome type counterparts and can be classified in a similar fashion. Chromatid interchanges can be classified on the basis of exchange (U or X), degree of completeness (complete or incomplete) and under certain conditions on the polarity of the chromosomes involved (Fig. 70).

Apart from their capacity to produce structural alterations both irradiation and chemical treatment can bring about direct alterations in chromosome number. This aspect of their influence has been far less studied than their structural effects. Cattanach (1964), for example, succeeded in inducing autosomal trisomy for one of the smallest autosomes (no. 18) in the mouse by TEM treatment of wild type hybrid (101 × C 3 H) males which were then mated to untreated females (see pg. 53). Likewise Griffen and Bunker (1964) produced viable trisomy for three different kinds of smaller chromosomes (nos. 8, 15, and 16) in the course of radiation studies in male mice (but see page 65).

2. Tissue Culture and Nuclear Transplantation

Normal tissues in primary culture or during early periods of *in vitro* life have predominantly normal chromosome complements. In this they agree with the striking constancy of chromosome number and morphology found within normal tissues *in vivo.* In contrast to this stability, the karyotypes of tumours, especially those that have been transplanted serially, and of permanent cell strains derived from non-neoplastic tissues may be highly irregular both in number and/or morphology. Three principal mechanisms are involved in the production of these abnormal chromosome complements:

(a) Doubling of the chromosome complement. This may occur in one of four ways:

(i) Endoreduplication—in which the chromosome number doubles during interphase,

(ii) Endomitosis—in which the chromosomes appear and double in number but without disappearance of the nuclear membrane, formation of spindle or division of the cytoplasm,

(iii) Failure of the spindle mechanism at mitosis, or

(iv) Production of binucleate cells following failure of cleavage.

The relative importance of these different mechanisms differs for different cell types but in all cases provided the resulting tetraploid is at no selective disadvantage it will persist in the cell population.

(b) Since they have double the quantity of genetic material, further chromosomal modifications can be tolerated or even selected for in such tetraploid cells. Thus in successive generations of serial cultures of embryonic mouse-skin cells, tetraploidy was followed by the development and establishment of predominating heteroploid stem lines (Levan and Biesele 1958). Such variation arises predominantly through non-disjunction.

A particularly interesting case of *in vitro* heteroploidy has recently been reported by Dolfini and Gottardi (1965) in *Drosophila melanogaster* (Table 55). In the first sample taken after the establishment of the culture (12 hrs.) 11.9% heteroploid cells were already present in the culture. The frequency increased to 29.7% at 18 hrs. and then remained constant for a further 30 hrs. During the interval between 48 and 72 hrs., however, the percentage of heteroploidy rose to 78.5% and then remained stable. Polyploid cells were virtually absent at all stages so that here, at least, heteroploidy is secondary and not a tertiary derivative via tetraploidy.

Table 55. *Percentage of Metaphases Showing Various Chromosome Numbers after in vitro Culture of Embryonic Cells of Drosophila melanogaster.*
84.4% of the abnormal cells involved chromosomes 1(X) and 1V while the two major autosomes (11 and 111) are involved in the remaining 15.6% of the abnormal cell types. Structural changes were also found.
(Data of Dolfini and Gottardi 1965.)

Duration of Culture (hrs.)	Chromosome Number													No. of Metaphases Analysed
	4	5	6	7	8	9	10	11	12	13	14	15	16	
12				3.0	88.1	5.2	3.0	0.7						135
18	1.6		1.6	6.3	71.9	7.8	4.7			1.6		4.7		64
21			1.7	10.3	79.3	1.7	3.4			1.7			1.7	58
24		1.3	2.5	7.5	72.5	10.0	3.8	1.3	1.3					80
48		1.6	1.6	8.2	70.5	6.6	3.3	6.6	1.6					61
72		12.3	4.6	29.2	24.6	5.4	7.7			3.1	3.1			65
96		5.0	11.7	26.7	20.0	5.0	6.7	5.0	6.7			1.7	1.7	60
120		10.0	3.3	38.3	20.0	20.0			1.7	3.3	3.3			60
	Hypodiploid (4–7)					Hyperdiploid (9–16)								

Hyper- and hypoploid cells have been reported also in diploid cell strains derived from the chinese hamster and the mouse and in clonal lines from pig kidney strain PK-Z (Chu 1961).

(c) Finally structural rearrangement of the chromosomes may also play a part in the repatterning of karyotypes *in vitro.* For example the normal complement of the mouse consists of 40 rod-shaped chromosomes which cannot be individually distinguished. In permanent cell strains, however, one of the commonest changes lies in the formation of metacentric chromosomes which contrast sharply with the uniform profile of rods observed *in vivo* and in primary culture.

The incidence of metacentrics varies between permanent cell lines and their origin has never been properly clarified, largely because the nature of the rod chromosomes in the mouse remains unresolved. One of the clearest cases so far analysed has been presented by OHNO, KOVACS, and KINOSITA (1960). Twelve distinct samples taken during the period November 1957 to April 1959 from L 4946 mouse ascites lymphoma all showed rod-type chromosomes though a proportion of hyper and polyploid cells was present. At the end of June 1959, however, a large mediocentric chromosome and one or more minutes appeared in the karyotype of hyperdiploid neoplastic cells. Variation in the number of minutes present reflects their mitotic instability. On the assumption that normal mouse chromosomes are acrocentric and that the minutes represented a short arm of one of a pair of acrocentrics OHNO *et al.* argued for a classical case of centric fusion. The validity of this argument, however, depends on mouse chromosomes being acrocentric and a strong case can be made for their centromeres lying in a strictly terminal position (see pg. 112).

The opposite change has been found in populations of chinese hamster cells (FORD and YERGANIAN 1958) and in human cells (CHU and GILES 1958). Here all the normal chromosomes have two arms and telocentrics arise in culture. Other types of morphological variation that have been detected in established cell lines include minutes, rings and less often dicentrics.

These various types of aberration appear to be determined by the *in vitro* environment which is quite different from their original *in vivo* habitat. Thus trypsinisation, used routinely to establish monolayer cultures, can produce chromosome aberrations (LEVAN and BIESELE 1958, CHU 1961). So too may the chelating agent versene (ethylenediamine tetracetate) which is often used for cell dissociation. Finally metabolites in the growth medium may also induce aberrations. Thus in asparagine-independent cell lines of JENSEN rat sarcoma some 40% of the cells showed one or more chromosome aberrations when cells were grown in a medium lacking asparagine. By contrast the original sarcoma cells grown in primary cultures containing asparagine showed no such effects. It is also worth bearing in mind that in some organisms, at least, even colchicine may produce structural rearrangements (e.g., *Collinsia*, SORIANO 1957).

Nuclear transplantation studies in amphibian embryos show that a change of cytoplasmic background also can induce changes in the chromosome complement. For example, HENNEN (1963) observed rings and fragments following the transplantation of diploid nuclei from *Rana pipiens* into experimentally enucleate eggs of *Rana sylvatica.* A comparable syndrome can be induced simply by injecting zygotes with homologous proteins from adult liver (URSPRUNG and MARKERT 1963).

3. Viruses and Malignancy

HAMPAR and ELLISON (1961, 1963) were the first to report the occurrence of alterations in the chromosome complement during the first post-infection mitosis of an aneuploid MCH-line of the chinese hamster following treatment with herpes simplex virus. Persistent chromosomal aberrations were subsequently found in clonal sublines derived from infected cells. A number of other workers have since found chromosome aberrations following infection of cells with simian vacuolating virus (SV_{40}), measles virus, chicken pox, polyoma and the Schmidt-Ruppin strain of Rous sarcoma virus which produces tumours in chickens and in a variety of rodents (Table 56).

Table 56. *Virus-induced Chromosome Breakage in Mammals.*

Species	Virus	Cell-type	Reference
1. Chinese hamster	Herpes simplex	In vitro heteroploid line	Hampar and Ellison 1961
		In vitro diploid line	Mazone and Yerganian 1963
	Measles virus	In vitro HEp-2 culture	Fjelde and Holterman 1963
	Adenovirus type 12		Stich *et al.* 1964
2. Rat	SR strain of Rous sarcoma	In vitro diploid rat embryo cells	Nichols 1963
3. Man	Clinical measles	In vivo peripheral leucocytes	Nichols *et al.* 1962
	Live attenuated measles vaccine		Nichols 1963
	Chicken pox		Aula 1963, 1965
	SR strain of Rous sarcoma	In vitro human leucocyte culture	Nichols *et al.* 1964
	Measles virus	In vitro heteroploid cell line LU 106 Human embryo kidney and adult peripheral leucocytes	Nichols *et al.* 1964, 1965
	Herpes zoster virus	Human embryo lung cells	Benyesh-Melnick *et al.* 1965
	SV_{40}	Diploid cells in culture	Moorehead and Saksela 1965
	Infectious hepatitis (presumed virus)	Leucocyte cultures treated with serum of infected patients	El-Alfi *et al.* 1965
	Aseptic meningitis (suspected virus)	Cultures of peripheral leucocytes	Makino *et al.* 1965

As far as the Rous sarcoma system is concerned the tumour shows a progression of chromosomal changes both *in vitro* and *in vivo*. This progression leads first to changes in the number of chromosomes around the diploid level. This is followed by changes in chromosome type and finally by changes in ploidy. In the measles system the disease was found to be associated with chromosome breakage in the peripheral white blood cells.

When patients with measles were studied by serial bleedings it was found that *for a short time only* after infection from 30–70% of the peripheral leucocytes contained chromosome breaks in contrast to the 0–5% usually seen in healthy control material (NICHOLS 1963). Similar breaks were found by AULA (1965) in clinical chicken pox. The timing of the samples appears to be critical because the periods of observed breakage are short, which may account for the failure of other workers to detect it (HARNDEN 1964). Finally in patients who had received live attenuated measles vaccine, breakage was again found, though to a lesser extent than in the disease

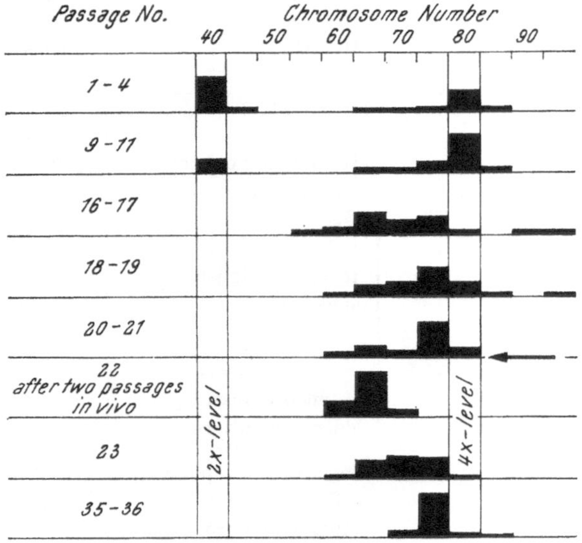

Fig. 71. Changes in the chromosome constitution of an embryonic mouse skin cell culture (after LEVAN and BIESELE 1958). The arrow marks the onset of malignancy.

itself. In both systems the abnormalities observed were primarily achromatic lesions (gaps) and open breaks. Breaks of the same type are known to be produced in human leucocytes by a group of ribosides which include FUdR, AdR, BUdR, araC and araA (NICHOLS 1966).

Notice, however, that the induction of breakage is not a universal property of all viruses. Thus the Bryan strain of the Rous virus will not produce breaks in human leucocytes although, as we have seen, the Schmidt-Ruppin Rous strain does so under the same conditions. Likewise virus-induced erythroleukemia cells, at both primary and secondary tumour sites, have the normal diploid macrochromosomes and probably a normal microchromosome content too. A normal chromosome constitution has also been found in two other virus tumour systems of the chicken and in seven different virus induced neoplastic conditions in mice and rats (BAYREUTHER 1959).

It is a fact of considerable interest that all the groups of agents known to produce chromosome breakage can also produce cancer. There is then the clear possibility that the potentiality for unrestricted proliferation characteristic of the malignant state might result from the unbalance produced by an irregular distribution of chromosomes. In 1958 LEVAN and BIESELE followed the chromosome constitution of a cell strain derived from embryonic mouse skin through successive transfers. Their results are shown in Fig. 71. In the first four passages normal diploid cells were well represented but their proportion decreased during the 9th—11th passages and they were replaced predominantly by tetraploid cells. Indeed, from the 16th passage on, the stem line was represented by hypo-tetraploid cells. After the 21st passage the cell strain proved malignant when tested *in vivo* and it developed into a tumour which proved to be transplantable. Here, unquestionably, we have a clear case of the *in vitro* origin of neoplastic transformation which appears to have been facilitated and enhanced, if not caused, by an abnormal chromosome constitution.

A particularly favourable material for the cytological study of malignant tissue is the free tumour cells in the ascitic fluid of mice, rats and men. Since 1950 there have been successive reports confirming that most ascites tumours are composed of populations of cells which vary from normal not only in their chromosome number but also frequently in their chromosome pattern. Moreover the range of variation can be altered experimentally by transplanting the tumour from one site to another. Variation in chromosome pattern may also be found between cells within a tumour. Cells with the commonest chromosome constitution form the "stemline" and the frequency of stem cells can vary and change during serial propagation of the tumour.

Established and fast-growing mammalian neoplastic cell strains propagated *in vitro* are just as abnormal and variable in their chromosome constitution as the tumours transplanted *in vivo*. According to HSU (1954) the neoplastic HeLa cell strain, derived from a human cervical carcinoma and propagated *in vitro* since 1952, is hypotetraploid, cells with 80—90 chromosomes being the most numerous. Secondary clones of HeLa origin show deviation in the frequency of stem cell types as well as numerical and structural changes in chromosome pattern (CHU and GILES 1958).

The conditions in tissue culture are known to be very different from those that exist in an organism (see pg. 135). Consequently it may be argued that the abnormal constitution of neoplastic cell strains is caused entirely by environmental factors. On the other hand it is possible that the primary explant itself contains cells with an abnormal chromosome constitution and that *in vitro* culturing favours these cells at the expense of cells with normal karyotype.

The first primary neoplasms to be examined cytologically were the human carcinomas. Here the conditions in peritoneal and pleural effusions are similar to those seen in the ascites tumours of mice and rats. Thus the cell population is heterogeneous in respect of both chromosome number and morphology. Moreover fluctuation both in the frequency of stem cells and in the drift of the stemline can occur. The cytological study of leukaemia, however, indicates a preponderance of cells with a diploid constitution. Indeed the evidence, as it stands, shows that many primary tumours are

Table 57. *Chromosome Constitution of Spontaneous and Induced Mammalian Tumours.* (Data of Bayreuther 1960.)

Tumour Origin	Species	Tumour Type	No. of Cases Examined	No. of Tumour Sites Studied				
				Total	Normal Chromosome Constitution	Cells Counted	Abnormal Chromosome Constitution	Cells Counted
1. Spontaneous primary tumours	Mouse	Lymphocytic leukaemia	16	36	32	493	4	50
		Mammary adeno-carcinoma	12	12	9	357	3	121
		Hepatoma	6	6	4	164	2	98
		Ovary tumour	4	4	4	176	—	—
		Uterus tumour	10	10	9	371	1	21
	Cattle	Lymphocytic leukaemia	2	2	2	68	—	—
	Man	Acute myeloic leukaemia	2	2	2	307	—	—
		Acute lymphatic leukaemia	3	3	3	424	—	—
		Chronic myeloic leukaemia	2	2	2	367	—	—
		Chronic lymphatic leukaemia	1	1	1	127	—	—
		Total	58	78	68	2854	10	290
2. Virus-induced tumours	Mouse and rat	Various, both primary and transfer	41	94	85	2754	9	347
3. Chemically-induced tumours	Mouse	9.10-dimethyl-1.2-benzanthracene induced lymphatic leukaemia (one transfer generation)	4	20	30	441	—	—
		Ethyl carbamate induced mammary and lung primary tumours	10	10	10	302	—	—

composed predominantly of diploid cells (Table 57). The exceptions are the tumours induced by chemical carcinogens at the site of innoculation. In these, as was first shown by WINGE (1931), the proportion of chromosomally abnormal cells can be very high.

The balance of evidence points to the conclusion that abnormal chromosome number is not a necessary condition for neoplastic transformation. Nevertheless chromosomally abnormal cells can be present in primary tumours and the frequency of these may increase during the progressive growth of the tumour. When such a tumour is transplanted into suitable hosts the chromosomally abnormal cells can further increase in number and may eventually replace the diploid cell complement.

In summary, it would appear that in primary tumours, especially when the cell populations are studied during the early stages, chromosome variation is either totally absent or else very sporadic. Conversely in transplanted tumours and tumour cell-strains the chromosome constitution and pattern can exhibit an extremely variable spectrum of diversity. The comparison between primary and transplanted neoplasms is especially instructive and shows that the extreme heterogeneity of cell populations in experimental tumours is facilitated partly by the high level of mitotic activity conferred on cells by the neoplastic state and partly by the "unfavourable" environment existing in malignant growth. Variation in chromosome number and structure thus occurs in tumour cells not because it matters more than in normal ones but because it matters less.

This is not to rule out the possibility that chromosome heteroploidy may not also promote neoplastic transformation by creating metabolic instability in the cell. As we have seen the early work of LEVAN and BIESELE suggests just such a possibility for here an *in vitro* change in a heteroploid "normal" cell strain was accompanied by its transformation into a neoplastic state. The fact that leukaemia is some ten times more common in trisomic mongols compared with normal humans points in the same direction. Similarly deletion of a part of the long arm of a G chromosome in man is associated with chronic myelogenous leukaemia (NOWELL and HUNGERFORD 1960). This unique structural abnormality, termed the Philadelphia chromosome (Ph′) has been found in almost all patients with this disorder. Nevertheless there is now a substantial body of evidence to show that, in the main, chromosome abnormalities are secondary phenomena which follow rather then precede malignant transformation.

What applies to animal tumours seems to hold also in plants. For example, in the process of culturing the haploid prothalli of *Pteridium aquilinum, in vitro,* occasional tumour-like cell masses are encountered which, when isolated, retain their characteristic appearance and continue to grow in a completely disorganised manner in separate cultures. Cytological studies on these tumours (PARTENAN *et al.* 1955) showed that initially all of the divisions ocurring in them were, as expected, haploid and that the chromosome complement was in all respects normal. In subsequent culture, however, the chromosome number in the dividing cells became higher, usually reaching a level around tetraploid due to endoreduplication.

Superimposed upon this polyploidisation was a variable but very prevalent aneuploidy within single cultures. Despite these nuclear changes the tumours appear to continue growing indefinitely. The gametophytes of *Osmunda cinnamonea,* the cinammon fern, have also given rise to tumours *in vitro* with a behaviour similar to that found in *Pteridium* (Partenan 1956).

VII. The Induction of Epigenetic Changes in the Karyotype

The most obvious and common phenotypic changes in the chromosomes are those in relation to condensation. Condensation cycles are particularly conspicuous in relation to the mechanical movements of chromosomes. The mechanisms which determine these cyclic changes are not known and opinion varies between the matrical school and that which seeks to explain the phenomenon in terms of the properties of the basic chromosome thread. While chromosome condensation is known to be affected by a number of internal conditions and artificial treatments e.g. X-rays, colchicine and cold treatment, only a few studies have been specifically concerned with the problem. For example, in keeping with hypothesis of Anderson (1965), polycationic compounds appear to induce condensation and maintain it where it already exists.

Chromosome size is partly a matter of contraction but, clearly, other factors are involved as well. And, as we have seen (Section II, page 24), this size can vary between cells within the individual even when differences in DNA value are not involved. Pierce (1937) claimed that in *Viola conspersa* the concentration of phosphate in the medium had an effect on chromosome size this being greater at higher phosphate concentrations. His results may be summarised as follows:

Solution	Chromosome Volume μ^3	Nuclear Diameter μ
Tap Water	1.92	5.5
Minus PO_4	2.41	5.3
Complete Medium	4.75	7.3
Excess PO_4	7.39	7.8

A more detailed study of this effect by Bennett and Rees (1967) has shown that while size differences are associated with parallel differences in dry mass, DNA changes are not involved. They also found a curious difference between rye and *Allium.* In both, larger chromosomes were determined by higher phosphate concentrations but while the effect was mainly due to a length increase in rye, chromosome width was affected in *Allium.* A nitrogen effect on chromosome size has also been claimed in several genera (Lutman 1934). Evans, Durrant and Rees (1966) find that there is 15% more DNA in the nuclei of large as opposed to small stable induced-genotrophs of flax i.e., plants transformed into new distinct heritable types by nutrient conditioning. Compared with control levels there was a fall in the DNA level of small and a rise in the level of large genotrophs. These differences are in the main located in the chromosomes and appear to be spread over several, perhaps all, the members of the complement.

Although coiling is most conspicuous in relation to nuclear division, it is also involved in the control of synthetic activity (see Section I, page 6). The most distinctive and specific reflection of differential behaviour associated with condensation cycles is seen in the puffing activity of polytene chromosomes. This is known to reflect the heterocatalytic activity of the genetic material at the level of gene transcription. In consequence it varies with genotype, cell type and stage of development. And the induction of these phenotypic changes has now been achieved experimentally. One of the most significant findings is that injections of the moulting hormone, ecdysone, induces both premature metamorphosis and the puff-pattern which characterises this morphogenetic transformation in untreated individuals (Clever 1961). Of parallel interest is the observation that puff-patterns can be modified, even to the extent of inducing novel puffs, by the transfer of nuclei into alien cytoplasm (Kroeger 1960).

Kroeger (1963) has claimed that the explantation of salivary glands of *Chironomus thummi* into saline may result in a reversion of puff activity to that characteristic of a more juvenile stage of development (rejuvenation) but the rejuvenated puffs are sometimes over-large. It was also claimed that by changing the Na^+/K^+ balance it is possible to activate a graded series of bands to give a puff sequence identical with that which marks the transition from adult larva to pre-pupa in the intact animal. Further, KCl is said to induce a particular puff in isolated glands of both *Chironomus tentans* and *Ch. thummi* while ecdysone induces the same puff in the intact tissue. Sodium chloride treatment on the other hand induces puffs which disappear when a high titre of ecdysone is applied (Lezzi 1966).

On the basis of such results it has been argued (Kroeger 1966) that the effect of ecdysone on the behaviour of the bands is not direct but mediated by a control system in the nuclear sap which is concerned with the Na^+/K^+ balance. The juvenile hormone may act via the same system but while ecdysone tends to increase the K^+ concentration, the juvenile hormone maintains the Na^+ level. In these terms rejuvenation would occur in media containing high relative concentrations of sodium.

The results obtained by Clever (1965) on the other hand, do not support this view of ecdysone action. He claims that K^+ ions do not have a specific effect on ecdysone-sensitive bands. Further, bands which respond to K^+ do so independently while ecdysone induces sequential effects. It would appear also that while certain bands have only short periods of competence during which they can respond to K^+ induction (just prior to the pupal moult) this does not hold for their reaction to ecdysone.

Reversible changes in puffing behaviour have been described following temperature shocks also. Thus, in *Drosophila busckii* band 2 L 8 is puffed in normal third instar larvae grown at 25° C while bands 2 L 14, 15 and 20 are not. But puffs do appear in these loci following a 30 minute exposure to 30° C while the normal puff progresses. On return to 25° C the induced puffs regress while the normal one reappears after an hour (Ritossa 1962).

It was also found that these heat effects could be mimicked by both sodium salicilate and DNP treatment.

Similarly, puff changes induced by temperature treatment in *Drosophila hydei* can be simulated by treatment with high K^+ ion concentrations. In addition to non-specific effects, Fedorff and Milkman (1964) have reported one puffing change in *D. melanogaster* which could be induced only by L-tryptophan. More specifically, a puff (68 D L 111) never observed in normal development, was induced in about an hour after transfer to Ringer's solution containing 0.03 M of L-tryptophan. Puffs can also be induced by anaerobiosis and RNase treatment. It is conceivable that the latter, by inhibiting protein synthesis, acts via some kind of end product control. But attempts to induce puffs with puromycin and similar antimetabolites have not been successful.

Puff formation and regression involve cycles of coiling and thread extension. It would appear that each puff arises from a single band although its uncoiling may affect adjacent regions by virtue of proximity. The allocycly is thus associated with a highly specific and localised control of gene transcription. But chromosomes show a higher order of integration than that implied in a string of genes and an adjustment of gene activity is frequently effected by a block reaction of coiling and uncoiling. Even less is understood about control at this level but some properties of the control system have been revealed by studies on chromosome rearrangements. Thus, as we have seen, one of the X chromosomes in somatic cells of female mice and other mammals becomes heteropycnotic and inactive. The other does not become over condensed and remains potentially active. Normally, paternally- and maternally-derived X chromosomes are affected in equal numbers of cells.

It would appear that autosomal segments translocated to an interstitial position in an otherwise complete X chromosome are obliged to behave in an X-like manner both in regard to allocycly and activity. This condition has been observed in the case of the Cattanach translocation in mouse. This translocation was unidirectional in that X material was not reciprocated to the autosome. In this case there was no discrimination between the normal X and the translocated X^A: they were inactivated in an equal number of cells.

The Searle translocation, however, is an interchange so that two new types of chromosome were produced by it, X^A and A^X. Consequently females heterozygous for this translocation have three chromosomes which contain X material namely, X, X^A and A^X. The behaviour of these chromosomes was as follows:

(a) The normal X alone was pycnotic in nearly 90% of the cells compared with 50% for both normal females and those heterozygous for the Cattanach translocation.

(b) The X^A alone was pycnotic in only about 10% of the cells, compared with 50% in the Cattanach translocation.

(c) In two of the ninety three cells studied, both the X^A and the A^X showed heteropycnosity but with this important difference. The X^A shows a whole chromosome effect with regard to pycnosity (compare X^A in Cattanach) but even on the rare occasions when the X material of A^X is pycnotic, its autosomal segment is normal.

It is possible, therefore, that only the centric region of the X is capable of effecting a pycnotic state in associated autosomal segments. But the study of a more extensive series of translocations in relation to the peculiar behaviour of the X chromosomes in male *Sciara* warns against this interpretation. Thus, the nucleolar function of a region closely linked to the X centromere of mouse may be the important factor (compare *Tribolium*) or the significant feature may be the relative amounts of autosomal and X material in the new chromosomes. In the case of the spreading found in relation to V-type position effects in *Drosophila*, on the other hand, chromocentral heterochromatin is implicated but even here nucleolar functions cannot be ignored.

Of course, in regard to differential behaviour within females, even normal X chromosomes do not behave uniformly throughout the mammals and ancient rearrangements may be involved (see Fig. 25). Thus, differential pycnosity may affect the whole chromosome (mice and men), a whole arm (golden hamster) or only a terminal segment of one arm (*Microtus agrestis*). In the special case of the creeping vole, where the female soma is XO (see page 39), differential behaviour cannot be expected within females or even between sexes. There is, however, differential pycnosity within the X chromosome and each of its arms, in both males and females. And while allocycly typifies a terminal region of one arm, it is found in an extensive pro-centric segment of the other.

VIII. The Principles of Karyotype Evolution

Studies on induced chromosome mutation are usually arranged to allow the aberrations to be observed at the division immediately following their induction or soon afterwards. For the most part, therefore, they are detected prior to their subjection to mechanical selection. Further, dosages are chosen which give comparatively high rates of mutation. Spontaneous aberrations, on the other hand, since they usually arise at a low rate, are rarely observed in the cell or even the individual of their origin. Rather they are seen in their derivatives and, consequently, after they have been mechanically tested at mitosis, meiosis or both. And the mechanically competent chromosome changes are but a small sample of the potential range. What is more, even if new karyotypes are capable of persisting through subsequent divisions they are still liable to elimination following genotypic selection.

The spectrum of aberrations thus progressively diminishes as one moves from the induced to the spontaneous and from the spontaneous which are mechanically efficient to those that are also genetically efficient. There is, therefore, a clear need to distinguish between the entire range of aberrations that can be identified in spontaneous or induced cases and the much smaller proportion of them that have a capacity for indefinite transmission.

The former, Ford (1964) has called primary structural change, the latter structural rearrangement. Primary structural change thus denotes a source of variation while structural rearrangement implies the incorporation of the change into a viable karyotype. Thus the study of the chromosome complement is not simply one of counting chromosomes or even evaluating shapes. These numbers and shapes afford a direct approach to evolutionary problems.

1. The Formation of Karyotypic Variants

While it is true that the frequency of spontaneous breakage is, in general, low, cases of extensive breakage are on record. Perhaps the most remarkable is that of *Elymus farctus* (= *Agropyron junceum*) reported by

Table 58. *The Frequency of Cells Showing Spontaneous Structural Changes in a Sample of 439 Metaphase Cells from a Mutant Individual of Elymus farctus.* (Data of Heneen 1963.)

Type of Aberration	No. of Affected Cells
1. Iso-locus breaks	124
2. Chromatid breaks with fragments held in their original position	59
3. Free chromatid fragments	12
4. Tri-radials	12
5. Quadri-radials	19
6. Monocentric rings	45
7. Dicentric rings	2
8. Dicentric chromosomes	71
9. Changes involving three or more chromosomes	4
10. New chromosome types	26
11. Acentric chromosome fragments and/or telocentrics	17
Total affected cells	391

Heneen (1963). This is a tetraploid species, $2n = 4x = 28$, and in one plant only, raised from seed collected from a small natural population, the total frequency of cells showing detectable aberrations was 74%. Other plants collected from the same locality were normal as too were collections from other localities. Aberrations and irregularities occurred in all the dividing tissues studied, which included root-tip cells, ovular tissues, tapetal cells and PMC's. Much, though not all, of the initial breakage in *Elymus* was of the chromatid type (Table 58).

The laws governing the occurrence of chromosome rearrangements depend on the properties of kinetochores and telomeres and on the nature of chromosome breakage. It has been believed for a long time that truely terminal chromosome deficiencies are not recoverable after irradiation in *Drosophila* (Muller 1932). This led Muller to propose the term "telomere" for natural terminal segments. It is admitted, of course, that the telomere concept may not apply in non-localised kinetic systems. Certainly the survival of fragments is far less hazardous. For example, EVANS and POND (1964) have shown that in *Luzula* there is little or no loss of irradiation induced fragments at mitosis. Moreover, as a result of the retention of kinetic activity by such fragments, micronuclei are rare following irradiation. The evaluation of the distribution of fragments between cells nevertheless leads to the conclusion that cells with relatively high numbers of fragments are selectively eliminated from the meristematic cell population. This, however, may depend on genic change rather than chromosome damage.

In a like manner the scorpion *Tityus bahiensis* which has no localised kinetochores does not have a constant chromosome number. In Piracicaba most individuals have 2n = 10 but at Ouro Preto 2n = 17–19. This suggests that fragmentations or fusions (or both) must have taken place during the history of the species. Indeed fragmentation-fusion heterozygotes occur in present day populations for metaphase chromosomes broken into two pieces have occasionally been found. In addition many of the individuals were interchange heterozygotes (Table 59). The size of the interchange associations ranged considerably and, curiously, included numerically uneven closed multiples. The process of transverse fragmentation inferred here and in *Luzula* (see pg. 127) presupposes that fracture sites automatically become stable ends.

Table 59. *Karyotype Variation in the Scorpion Tityus bahiensis.* (Data of DA CUNHA and PAVAN 1954.)

2n	Meiotic Karyotype
5	1.II + 1.III 2.II + 1.I
6	3.II 2.II + heteromorphic. II 1.II + 1.IV 1.VI
7	1.VII
8	2.II + 1.IV
9	1.II + 1.VII 1.IX 1.III + 3.II
10	5.II 2.III + 2.II 1.II + 2.IV 1.IV + 3.II 1.VI + 2.II
17	1.XVII
18	1.XVIII
19	1.XIX

Observations contrary to the telomere concept are known also in localised kinetic systems. Thus in *Zea mays* breakage of a dicentric at meiotic anaphase gave rise to a chromosome 9 whose short arm terminated in a broken end. During the ensuing gametophytic division sister union occurred at the position of the breakage and resulted in a repetitive breakage-fusion-bridge cycle and a persistent dicentric (MCCLINTOCK 1941). When this dicentric is introduced into the endosperm tissue through either the male or the female gametophyte the breakage-fusion-bridge cycle continues to operate. But if introduced into the zygote the cycle discontinues, the broken ends heal and the healing is permanent. The broken ends now behave in every respect like normal ends. Thus the breakage-fusion-bridge cycle occurs only in the nuclear divisions of the gametophytic and endosperm tissues and even then only provided the broken ends are newly derived and have not passed through a sporophytic generation.

The factors responsible for fusions of broken ends or for their healing are probably related to the method by which the chromosome becomes broken and to the physical conditions in the vicinity of the broken end. The broken chromosomes studied by McClintock all originated as a consequence of mechanical rupture at ana-telophase. Breakage induced by other means need not lead to similar consequences. Again Fabergé (1959) found that about 35% of all the pro-terminal breaks induced in maize pollen by α-particles were stable whereas X-rays or UV-particles produced few, if any, stable ends in the same material.

If telomeres are natural chromosome elements which can arise only from other telomeres then it is difficult to explain their *de novo* production in maize. Fabergé suggests that the α-particle track, unlike X-rays, caps the broken ends temporarily with decomposed material which prevents immediate or rapid sister fusion. With such provisional protection the broken ends may be able to acquire a natural "coating" or to generate a new telomere which makes them non-sticky. The "telomere" concept is thus not of absolute validity—changes can, as indeed Muller and Herskowitz (1954) realised, occur in polarity. Especially pertinent is the process of misdivision where the terminal structures produced by the fracture of the initial single kinetochore of the metacentric element must become both kinetochores and telomeres. Perhaps the distinction between stable and unstable telocentrics depends upon the capacity of the terminal kinetochore so produced to function as an effective telomere.

Be this as it may the general principles governing the rearrangement of chromosomes have been formulated with a belief in the telomere concept and are three in number:

(i) Freshly broken surfaces tend to fuse with other broken ends or their replication products whereas natural ends show no tendency to fuse with either freshly broken ends or with natural ends,

(ii) Chromosomes with broken ends are not stable and will not survive in the course of successive divisions. Two or more breaks are thus necessary to secure viable structural rearrangements, and

(iii) Centromeres are self-reproducing organelles which cannot arise *de novo* nor lose their properties.

If these rules are accepted then breakage without reunion does not give rise to stable alterations of the karyotype while reunion without previous breakage cannot occur at all. However, as we have already seen, there is evidence contrary to (ii) both in species with and without localised centromeres. Likewise (iii) omits any consideration of the fact that centromeres are compound structures and hence are capable of subdivision into simpler but still functional units.

Superimposed upon these primary rules are some secondary ones for there is now unequivocal evidence that in some species certain chromosome regions are more or less likely to engage in structural change than others. Two such regions are of particular importance—the heterochromatic regions and the nucleolar organisers. The influence of these organelles has been especially well demonstrated in *Vicia faba* by Evans and Bigger (1961).

These authors showed that with respect to irradiation-induced interchanges and isochromatid aberrations there was no direct proportionality to length. Thus by dividing the total complement into two groups, M and S (Fig. 72), the length ratio at metaphase is M/S = 1/2.16. But the frequency with which M and S chromosomes undergo interchange both in 2x and colchicine-induced 4x cells is 1/3.44, a ratio which differs significantly from that expected on the basis of chromosome length. The comparable ratio for iso-chromatid aberrations is even higher—1/4.4.

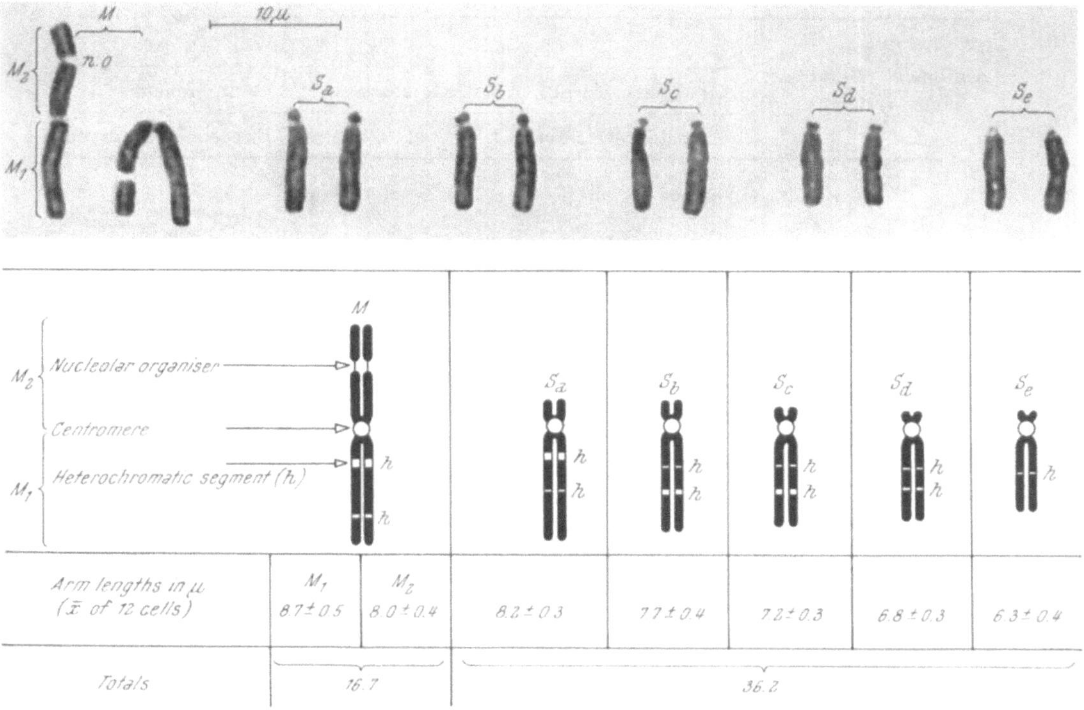

Fig. 72. Characteristics of the chromosome complement in *Vicia faba* (2n = 12). The positions of heterochromatic segments are indicated by clear regions (h). Data of EVANS and BIGGER 1961.

A more detailed analysis revealed the interesting facts that:

(i) As far as interchanges were concerned the nucleolar arm (M_2) showed a larger deviation (reduction) from expectation than the M_1 arm, while amongst the acrocentric S-chromosomes the bulk of the deviation is due to the shortest chromosome (Se) which participates in interchange more frequently than expected (Table 60).

(ii) For the iso-chromatid aberrations both arms of the M-chromosome undergo iso-chromatid intrachange less often than expected but here the M_1-arm shows a larger deviation than the nucleolar arm. Again the bulk of the deviation in the S-chromosomes is due to S_d and S_e, both of which have approximately twice as many iso-chromatid breaks as expected on the basis of metaphase length.

By accurately mapping the location of the points of breakage and exchange it was shown that:

(i) The principal deviation from randomness in the S-chromosomes is due to a marked deficiency of interchanges and iso-chromatid aberrations

Table 60. *Distribution of Structural Changes Induced by* $Co^{60}\gamma$*-irradiation of the Chromosome Complement in Vicia faba.*

The 4x cells are from roots treated with 0.05% aqueous colchicine for 3½ hrs. prior to irradiation. The expectations are those based on metaphase length.
(Data of EVANS and BIGGER 1960.)

Chromosome	Metaphase Length (μ)	2x-cells				4x-cells	
		Iso-chromatid bks.		Interchanges		Interchanges	
		Expected	Observed	Expected	Observed	Expected	Observed
M_1	8.7	33	13	66	57	33	31
M_2	8.0	30	24	60	42	30	18
Sa	8.2	31	27	62	53	31	39
Sb + Sc	14.9	56	46	112	107	56	50
Sd	6.8	26	47	52	78	26	29
Se	6.3	24	43	48	63	24	33
Totals	52.9	200	200	400	400	200	200

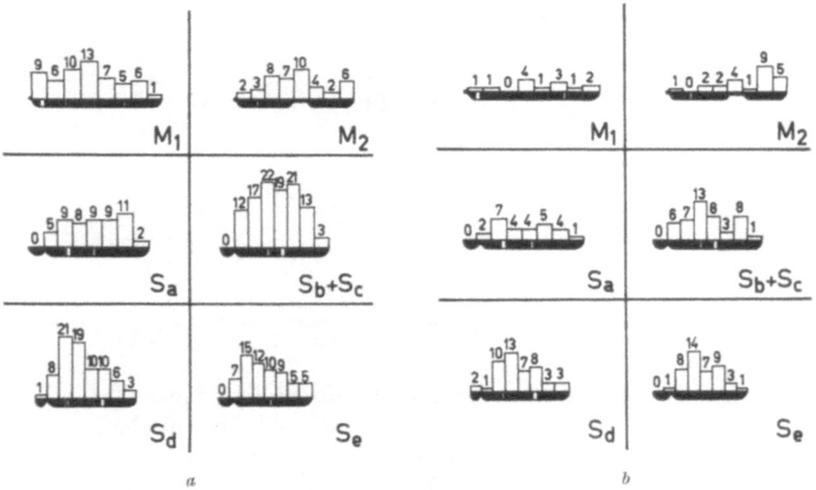

Fig. 73. Frequency distributions of 200 interchanges (*a*) and 200 isochromatid aberrations (*b*) in X-irradiated diploid cells of *V. faba* (after EVANS 1961).

near the centromere and the ends of the chromosomes and an excess of aberrations in the mid regions (Fig. 73).

(ii) In the M-chromosomes the distribution of interchanges is not significantly different from randomness but the iso-chromatid data indicate a significant deviation. This is due to an excess of aberrations in the satellite of the M_2 arm.

Thus the difference in distribution of H-material between the M and S chromosomes parallels, and would appear to be the cause of, the difference in the distribution of interchanges within the two chromosome classes. H-regions appear as Feulgen-positive heteropycnotic segments in the interphase nucleus and although as many as 20 discrete H-regions may be observed in *Vicia faba* they usually fuse to give a smaller number of larger chromocentres. The H-regions and hence the adjacent euchromatic regions are thus frequently in contact or at least in close proximity. And this proximity clearly facilitates exchange both within and between chromosomes at these points. Notice, however, that although the excess of aberrations in the mid regions may be explicable in terms of the distribution of heterochromatin, this explanation cannot be applied to the deficiency of aberrations at the chromosome ends—indeed the cause of this deficiency remains unresolved.

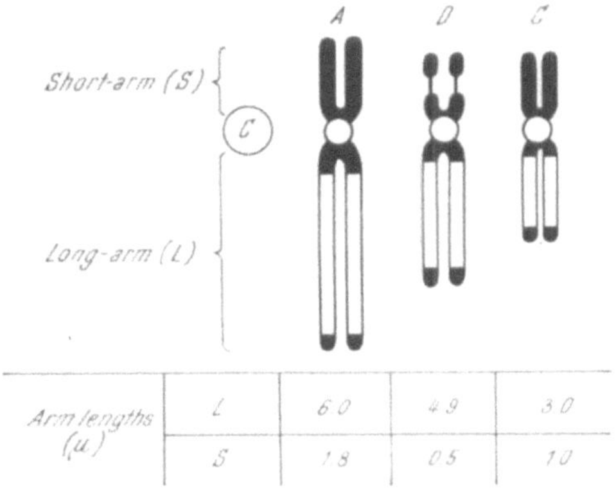

Fig. 74. Characteristics of the chromosome complement in *Crepis capillaris* (2n = 6). The positions of heterochromatic segments are indicated by solid regions (data of SIRE and NILAN 1959).

A rather different explanation can be applied to the reduced frequency of interchange in the M_2 nucleolar arm. In the interphase nucleus the nucleoli occupy about 20% of the total nuclear volume for each nucleolus is about 7μ in diameter. Clearly, then, the nucleoli attached to the M_2-arm must considerably reduce the probability of close proximity or contact between the M_2 and any other component of the complement.

Two other examples demonstrate the importance of the amount and distribution of H-material. Two species of *Impatiens* are known with roughly equal amounts of heterochromatin. In the one (*I. sultani*, $2n = 16$) there are on average five H-regions per chromosome while in the other (*I. balsamina*, $2n = 14$) there is only one H-segment per chromosome. Significantly the former shows a much higher frequency of X-ray induced aberrations than the latter (BHATTACHARJYA 1958). Likewise in *Crepis capillaris* ($2n = 6$) there are three pairs of chromosomes—A, C and D—whose length and H-properties are summarised in Fig. 74. SIRE and NILAN (1959) report a marked departure from randomness in the breakage response to X-rays of the components of this complement. Since the long arm of the A-chromosome is the longest in the complement it might have been expected to have more dicentric aberrations than the long arms of C or D. But this was not the case (Table 61). Breaks leading to dicentrics occurred with equal frequency in the long arms of all three chromosomes. Thus, although the breaks per chromosome arm were equal, breaks per unit length were not. However, when compared with the amounts of heterochromatin, the observed breaks equalled the expected. In the short arms, with the exception of that of D—the nucleolar organising chromosome—the frequencies of dicentric exchanges were proportionate to length. Significantly the short arms in *Crepis* are entirely heterochromatic. The data pertaining to simple breaks are less extensive but, in general, fit the same scheme (Table 62).

That simple breaks are most frequent in euchromatic regions argues for increased rejoining leading to dicentric formation rather than increased

Table 61. *Distribution of Breaks Leading to Dicentrics Following X-irradiation of the Chromosome Complement in Crepis capillaris.*
(Data of SIRE and NILAN 1959.)

Chromosome Region		Observed	Expected on Basis of Total Length	$\chi^2_{(1)}$	Expected on Basis of Het-length	$\chi^2_{(1)}$
Long Arm	A	198	259	15.00**	196	0.020
	C	209	130	44.42**	196	0.863
	D	182	212	4.53*	196	1.000
Short Arm	A	81	78	0.05	90	0.900
	C	48	43	0.21	50	0.080
	D	36	22	6.50*	25	4.840*

Table 62. *Distribution of Simple Breaks Following X-irradiation of the Chromosome complement in Crepis capillaris.*
(Data of SIRE and NILAN 1959.)

Simple Breaks	Chromosome Arm					
	A		C		D	
	Long	Short	Long	Short	Long	Short
1. Observed	77	13	38	0	21	0
2. Expected on the basis of total length	52	16	42	4	26	9
$\chi^2_{(1)}$	12.02**	0.56	0.38	4.0*	0.96	9.0*

breakage. Thus of 200 dicentric exchanges analysed in detail with regard to position, 65% were located definitely in H-regions, 2% were in euchromatin and the remaining 33% could have been in either. On the other hand of 100 simple breaks scored for location 62% were definitely in euchromatin, 24% were in heterochromatin and 14% could not be allocated.

That equivalent structural limitations operate for spontaneous as well as induced aberrations is indicated by observations of three kinds. First, naturally selected interchanges in *Blaberus discoidalis* (JOHN and LEWIS 1959) and in *Lycopersicon* and *Oenothera* are confined to heteropycnotic regions. Second, in *Anthoxanthum odoratum, Secale montanum,* and *Alopecurus pratensis* chromosome breakage is frequent at the secondary constriction (MARKARYAN and SCHULTZ-SCHAFFER 1958).

Thirdly, in man all 5 pairs of acrocentric autosomes (13–15 and 21–22) appear to carry nucleolar organisers on their short arms. These tend to co-operate in forming a common nucleolus rather than functioning independently. OHNO *et al.* (1961) suggest that the close proximity of these chromosomes, conditional upon this co-operation, may facilitate exchanges between them. Significantly the most common translocations known in the human complement involve the short arms of the members belonging to these two groups (see also pg. 12).

Table 63. *Influence of B-chromosomes on A-chromosome Stability in the Endosperm of Trillium grandiflorum.*
(Data of RUTISHAUSER 1960.)

No. of B-chromosomes in Endosperm	No. of A-chromosomes Analysed	No. of A-chromosome Fragments		$\chi^2_{(1)}$
		Total	%	
0	15,626	125	0.80	
1	8,182	106	1.30	13.50, P < 0.001
2	8,766	114	1.30	14.38, P < 0.001
3	780	10	1.28	2.81, P > 0.05

Table 64. *Influence of B-chromosomes on A-chromosome Stability in the PMC's of Crepis capillaris.*
(Data of RUTISHAUSER 1960.)

No. of B-chromosomes	No. Micronuclei per Tetrad
0	0.016
1	0.13
2	0.36

One final factor concerned with karyotypic plasticity is genic background. Several facts support the assumption that spontaneous chromosome aberrations are subject to genotypic control:

(i) A type of chromosome breakage in inbred rye has been shown by REES and THOMPSON (1955) to be genetically determined. And crossing experiments indicate that the abnormality is controlled by more than one gene.

(ii) In *Trillium grandiflorum* (Table 63) and *Crepis capillaris* (Table 64) B-chromosomes raise the level of chromosome breakage. In *Myrmeleotettix maculatus* supernumerary chromosomes not only increase breakage frequency but also increase the frequency of production of polysomics (HEWITT and JOHN 1965). Supernumerary segments in *Chorthipus parallelus* may lead to extensive polysomic mosaicism for the M_4 member of the A-set (Fig. 75) and SANNOMIYA (1962) has found some evidence for the same process in B-containing males of the grasshopper *Patanga japonica.* Indeed there is evidence that in some organisms a certain amount of heterochromatic

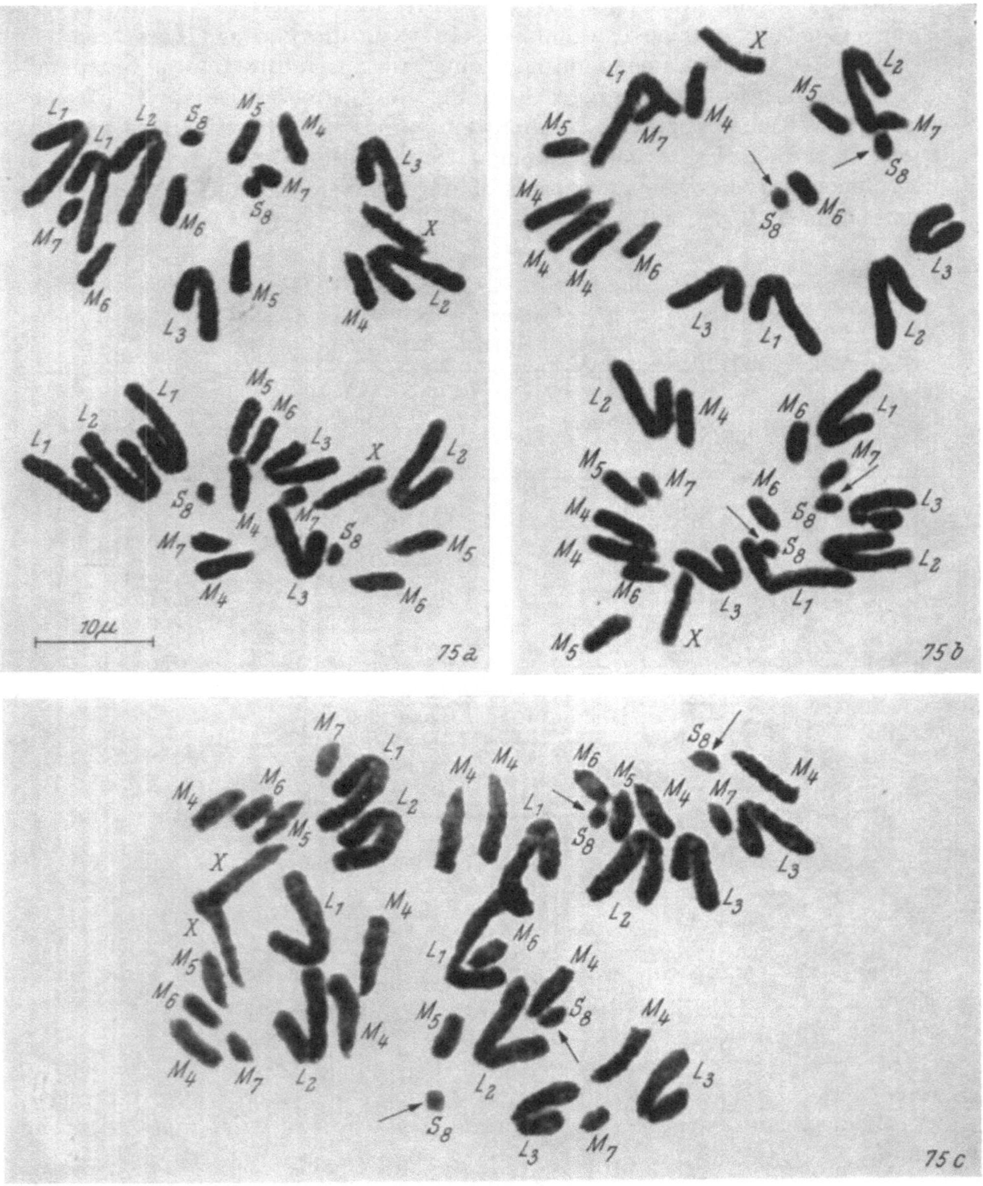

Fig. 75. Instability of the male germ line mitoses in the presence of S_8 chromosomes heterozygous for a supernumerary chromosome segment in *Chorthippus parallelus* (see Fig. 45).
a — Normal diploid anaphase (2n = 17), S_8 in the form of the basic homozygote, *b* and c — Cells respectively tetrasomic (2n = 19) and pentasomic (2n = 20) for the M_4 chromosome; in both cases the S_8 is present in the form of the structural heterozygote. Note that the extra M_4 chromosomes tend to appear slightly smaller than the standard M_4's, a consequence of their increased chromaticity (see also Fig. 81).

material may be essential in creating a state of karyotypic plasticity. For example in the american species of *Tradescantia* there is no detectable heterochromatin. Chromosome mutants of normal range are formed spontaneously at quite high rates (GILES 1940, 1941). Despite this the only structural changes encountered in the american species of this genus in nature are inversions (SWANSON 1940) and even their incidence may have been exaggerated (LEWIS and JOHN 1966).

(iii) The importance of heterochromatin is seen also in *Chilocorus stigma* (SMITH 1959). Fundamentally the diploid number of this species is 26. Decreases in number have been brought about by centric fusion, increases by the formation of B-chromosomes or supernumerary chromosome arms

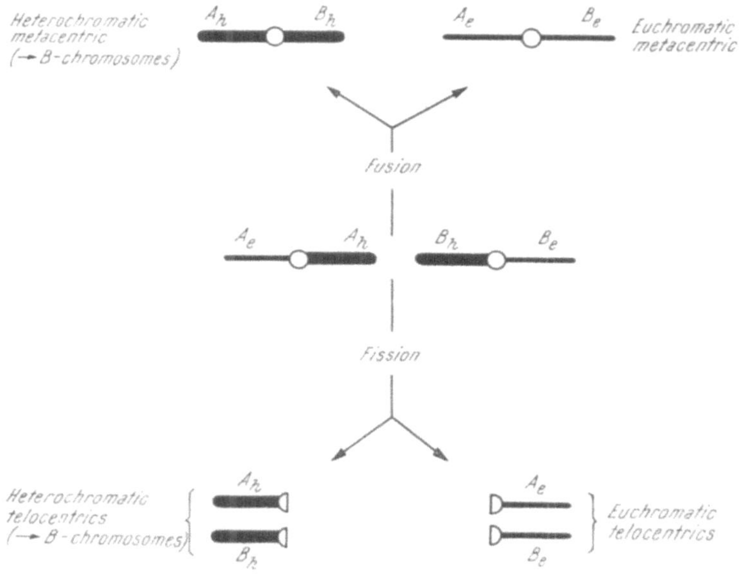

Fig. 76. Chromosome relationships in the coleopteran genus *Chilocorus* (based on SMITH 1959).

which float in the species in variable numbers. Changes in shape arise from complete loss of, or partial loss from, one arm of the autosomes. All these variables contributing to the total polymorphism are rendered physiologically and genetically acceptable because of the restriction of the epigenetically active euchromatic material to one of the two equal sized arms of each initial chromosome, the other being totally heterochromatic (Fig. 76).

(iv) GERSTEL and BURNS (1965) studied triploid hybrids between *Nicotiana otophora* (2 n = 24) and *N. tabacum* (2 n = 48) and also derivatives of the hexaploid hybrid induced by colchicine. The majority of the cells in the derived lines had the expected chromosome complement. In certain individuals, however, cells with fragments and dicentric or ring chromosomes were found and numerical variation was also apparent. These individuals also contained some remarkable "megachromosomes" which were as much as 15 times the length of the longest chromosomes in the normal complement. They tended also to be slightly thicker but variation in size and form was apparent even within individuals. Megachromosomes occurred only in a small minority of cells and these were almost invariably scattered. It would appear, therefore, that they are not transmitted mitotically. Nevertheless they reappeared in consecutive generations of several lines. This, clearly, does not depend on the transmission of the megachromosomes themselves but on the inheritance of the ability to generate them. The nature of these megachromosomes is obscure and their mode of origin has not been elucidated.

(v) LEVITAN (1964) has recorded over 2,000 new chromosome aberrations in *Drosophila robusta* over a period of six years. These were produced through the mediation of an inherited maternal, apparently cytoplasmic, factor in one of his homokaryous stocks (STy) which leads to random chromosome breakage.

2. The Stabilisation of Novel Karyotypes

Particular kinds of karyotypes sometimes characterise major taxonomic groups. This depends not only on the laws which govern the actual occurrence of rearrangements and changes but also on the principles which determine the probability of the various permissable types of rearrangements actually establishing themselves in natural populations. This probability, in turn, depends on factors of two kinds—mechanical and genetical.

The mechanical requirements for successful mitosis and meiosis include the co-adaptation of chromosomal dimensions to spindle mechanisms and the activity and interaction of the centromeres on the spindle. Thus, as SWANSON (1957) points out there must be an upper limit to the size of a chromosome. This limit is determined by the maximum distance from the pole of the spindle to the metaphase plate. Chromosomes longer than this would presumably suffer loss of their terminal regions when the cytoplasm constricts at the end of division. Limits exist also in relation to the number of chromosomes which a given system can support functionally. For example, spindle defects in *Achillea* (Fig. 77) are more common in 4 x and 6–8 x individuals than in diploids (EHRENDORFER 1959). Likewise when one B-chromosome is present in rye, mitosis is normal but with three B's the centromere-spindle mechanism is upset and when two are present the effect is intermediate.

Again, the great majority of pericentric inversions which have established themselves in grasshopper populations have done so in chromosomes that were originally rod-shaped. WHITE (1961) argues that this process tends to decrease the inter-centromeric distance in the meiotic bivalent and that this increases the "safety-factor" determining co-orientation on the spindle. In support of this argument WHITE found frequent malorientation in a pericentric inversion of the AB-metacentric in *Moraba scurra.* Indeed it would appear that it is this malorientation rather than chiasma formation in the reversed loop which has been the principal factor preventing the establishment of pericentric inversions in the AB-chromosome of morabine grasshoppers for this chromosome never shows a balanced polymorphism for pericentric inversion. Where on the other hand it has become "dissociated" into two rod elements, A and B, one or other of these is frequently polymorphic for a pericentric inversion. One cannot, however, generalise from this example because co-orientation may be impaired or delayed not only because the effective inter-centric distance is too long (UPCOTT 1939, KLINGSTEDT 1939) but also because it is not long enough (REVELL 1947, REES 1955). The location of the chiasmata will also affect this distance (LEWIS 1958).

In the midge *Tendipes decorus* 54 of the 70 known paracentric inversions occur in the second chromosome within which inversion breaks are again clustered (ROTHFELS and FAIRLIE 1956). Successive inversion steps frequently show near-coincidence of breaks and sometimes even exact

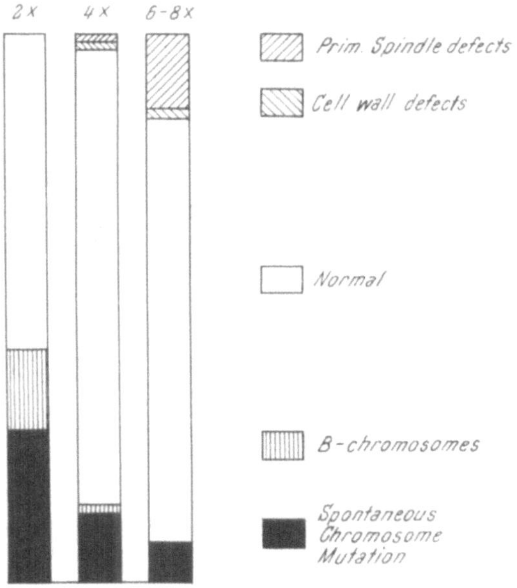

Fig. 77. The relationship between the level of ploidy and the percentage of spindle defects in the development of the pollen in the *Achillea millefolium* complex. Note also the correlation between the incidence of spontaneous chromosome aberrations and the level of B-chromosomes in the various ploidy types (after EHRENDORFER 1959).

coincidences so that near tandem and tandem complexes result. Here, therefore, existing inversion heterozygosity facilitates the formation of derivative inversions related by one or more breaks to the antecedant.

Finally, the analysis of species hybrids (*Eyprepocnemis*, JOHN and LEWIS 1965) or the comparison of the karyotypes of related species (Truxaline grasshoppers and locusts, JOHN and HEWITT 1967) reveals the existence of structural changes which bring about reorganisation of the karyotype without changing its general morphology. This is best seen in the genus *Chironomus* where 21 of the species studied cytologically by KEYL (1962) fall into five complexes which are related to one another by reciprocal translocations of entire arms (Figs. 78 and 79). Superimposed on this are paracentric inversions in the A, E and F arms. In a parallel fashion the H-material in all species of *Trillium* is concentrated in the same regions—in the short arms of the B and D chromosomes and proximally in the long arms of C and E (Fig. 80). Thus, the variation in the H-segments follows a similar pattern in different species (DARLINGTON and SHAW 1959). Indeed the polymorphism is parallel not only in each chromosome but in each arm of a chromosome. For example the two arms of A are always similar, those of B are always widely different. There must, therefore, be principles limiting the distribution of heterochromatin.

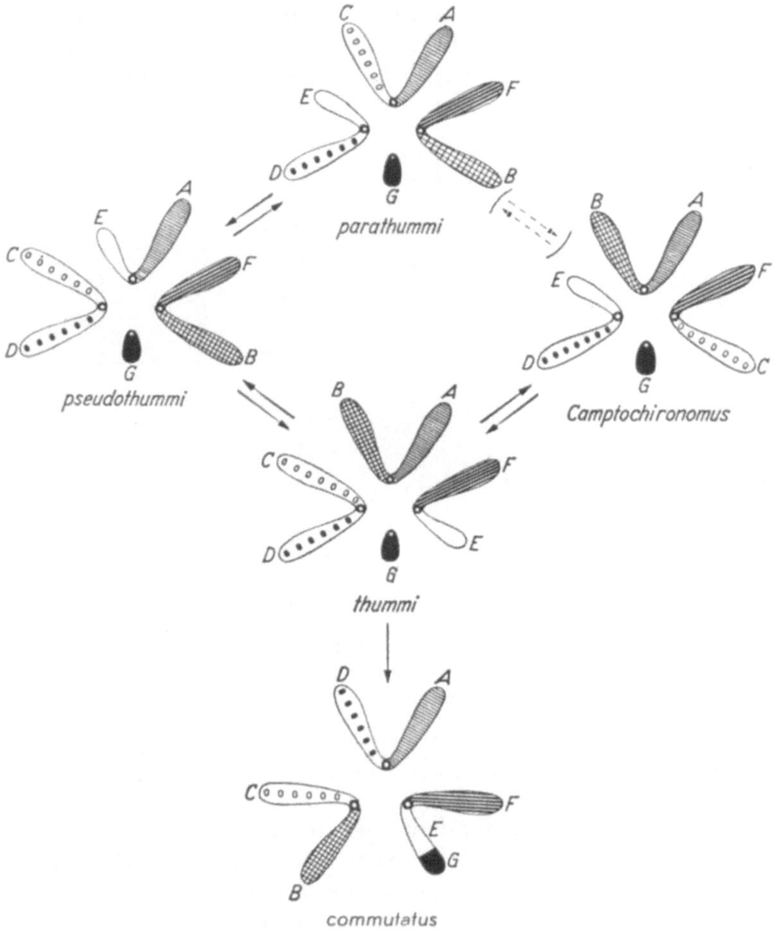

Fig. 78. Karyotype inter-relationships in the genus *Chironomus* (after KEYL 1962).

Genic as opposed to mechanical requirements are also clearly involved in certain cases. Thus in 7 of the 10 induced X/A translocations so far studied in the mouse the male carriers have turned out to be sterile since the primary spermatocytes degenerate in pachytene. CATTANACH's (1961) TEM-induced translocation is fertile but this is an insertion and not an interchange (OHNO and CATTANACH 1962). In the case of the T (X; ?) 16 H translocation (FORD and EVANS 1964) four younger animals (6–9 weeks old) proved to be normal at diakinesis and first metaphase but showed no stages beyond metaphase-II and no mature sperm. In two older animals (19 and 31 weeks) only sertoli cells and a few scattered spermatogonia were found.

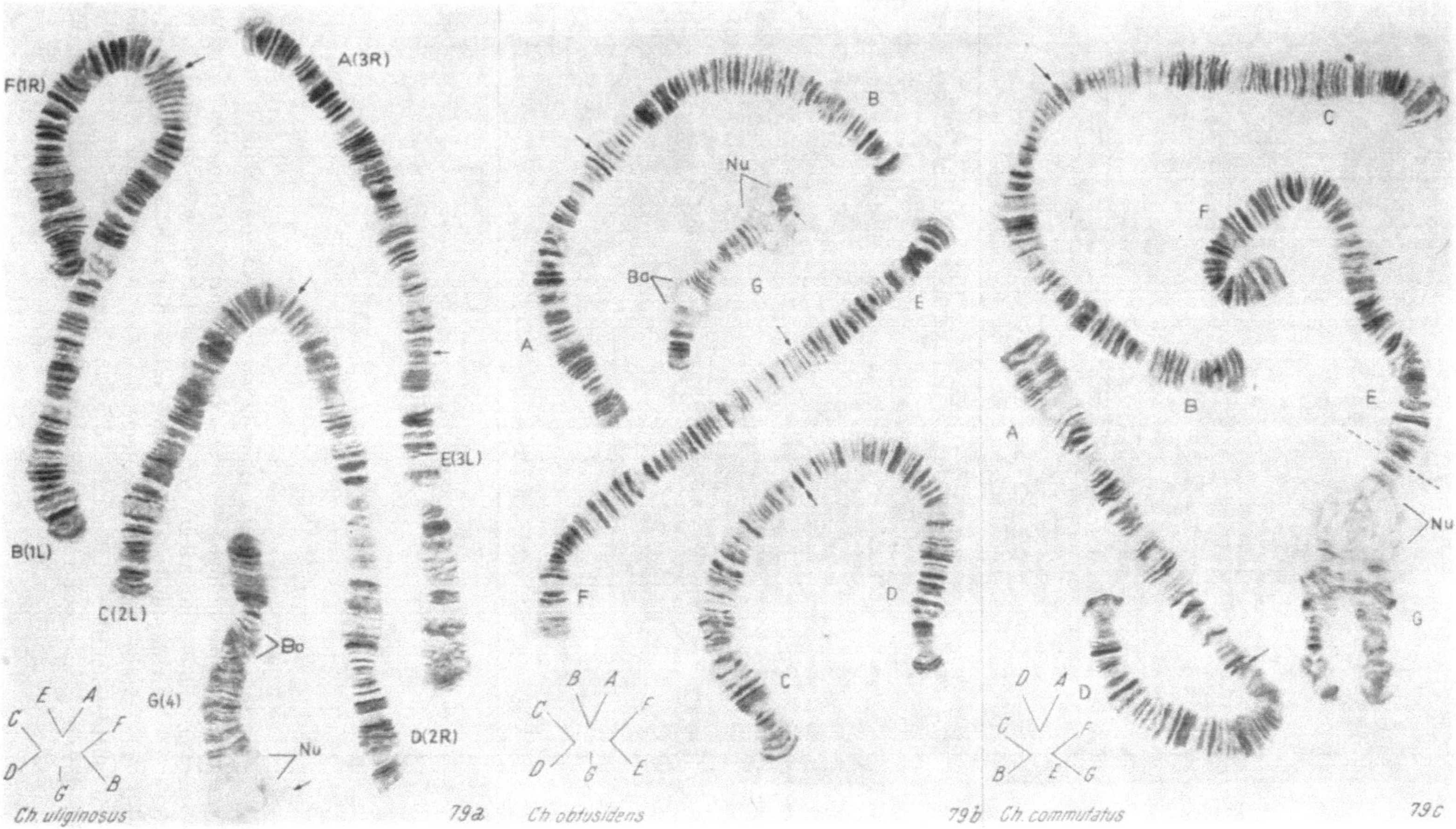

Fig. 79. Polytene chromosomes in the genus *Chironomus* illustrative of the evolutionary patterns summarised in Fig. 78 (after KEYL 1960).

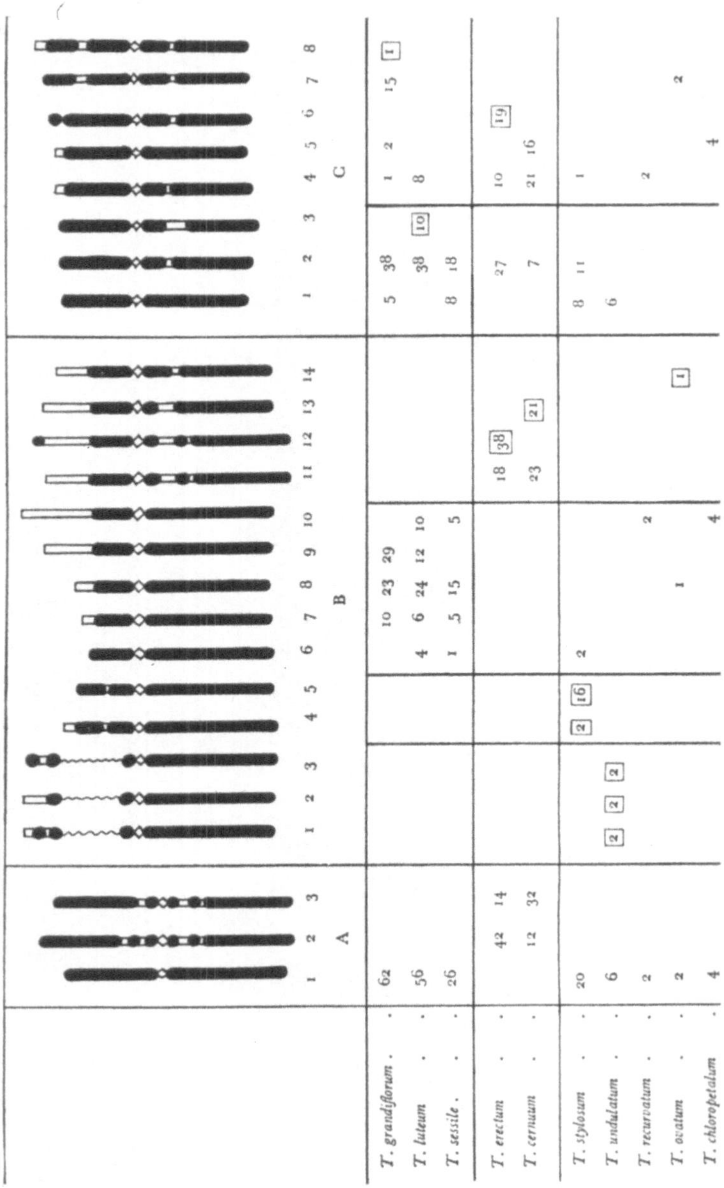

Likewise in the sterile translocation heterozygotes studied by LYON and MEREDITH (1966) spermatogenesis was arrested during first meiosis. Again the trisomic mouse described by CATTANACH (1965) was detected only because it was completely sterile since otherwise it was phenotypically normal. The vitality of this trisomic must have been good for the animal was in perfect health when sacrificed at the age of 8 months. The testes were normal in size but few spermatogenic cells were found beyond the

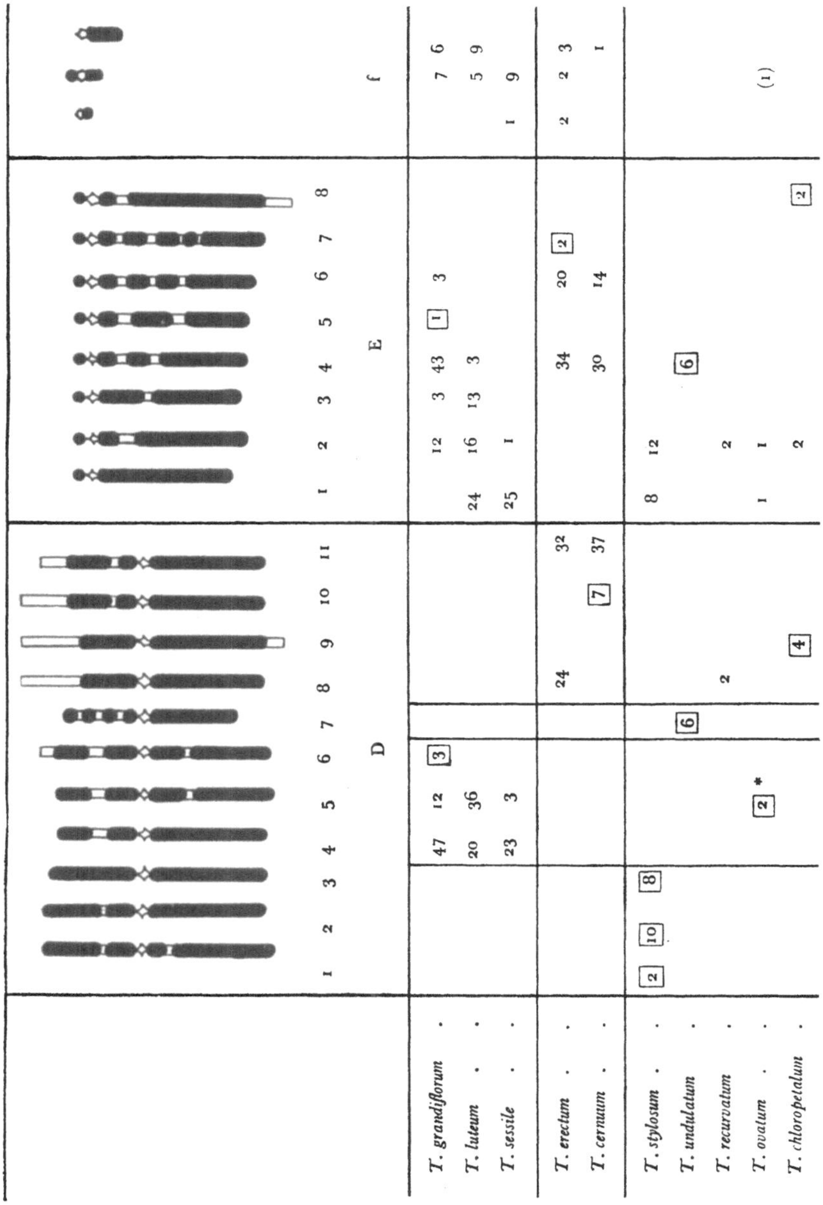

Fig. 80. Patterns of polymorphism for cold-induced (1–2° C for 96 hrs.) H-segments in ten species of the genus *Trillium*. The five pairs of chromosomes which characterise the diploid complement of all ten species are designated according to the european system as A—E. The american and japanese equivalents of the european system are as follows:

European	A	B	C	D	E
American	E	B	D	C	A
Japanese	A	C	B	E	D

Each variant of a particular chromosome type is given a number from 1 to 14. The numbers in squares refer to chromosome types which are peculiar to one species. Finally 'f' refers to the supernumerary chromosomes which characterise certain of the species. Note that the position of the centromere is denoted by a diamond while the het-segments are represented as unblocked regions (after DARLINGTON and SHAW 1959).

first meiotic division. Spermatids and spermatozoa, likewise, were all but absent and most were abnormal in shape. The sterility of this trisomic thus resulted from spermatogenesis breaking down after the first meiotic division. Again in *Campanula persicifolia* interchange homozygotes are never viable (DARLINGTON and LA COUR 1950) either in selfed or crossed progeny. Presumably, therefore, the breakage sites are either linked to recessive lethals or the aberration itself must be regarded in this way.

In cases such as these we are faced with the problem of deciding why the structural rearrangement causes death, sub-vitality or sterility. One thing appears to be clear. The sterility in translocation mice can hardly be attributed to segregation because many of them fail before anaphase and

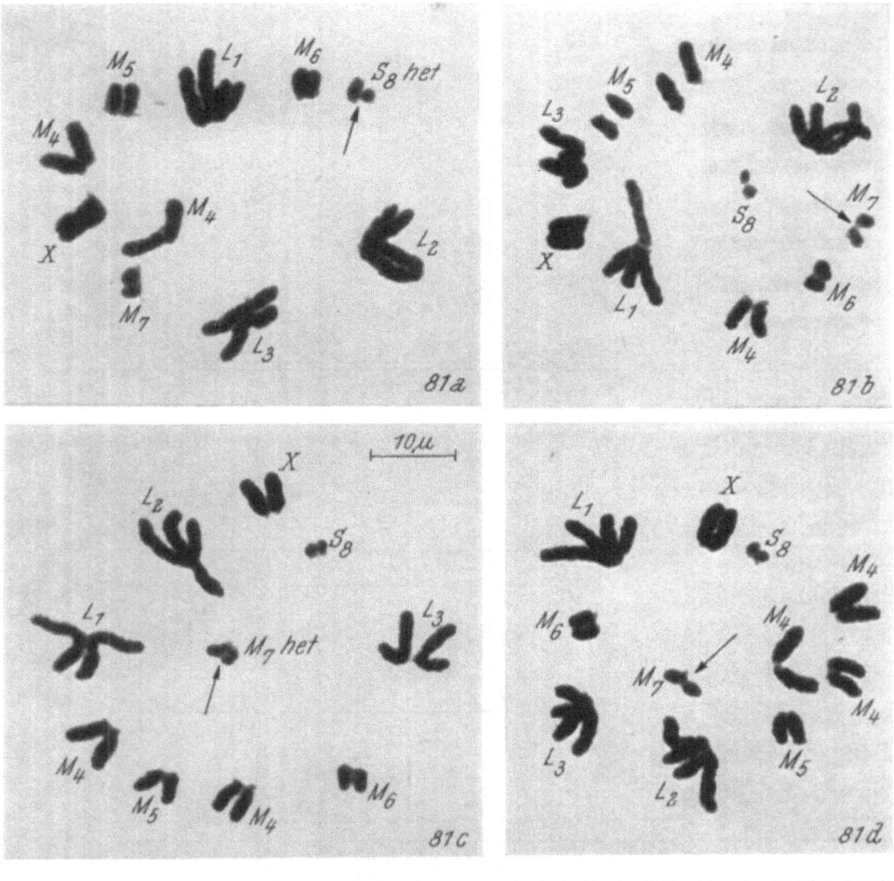

Fig. 81. Second metaphase plates with an extra M_4 chromosome in individuals of *Chorthippus parallelus* carrying supernumerary segments on the S_8 or M_7 members.

a — heterozygous S_8
b — structurally homozygous M_7
c — heterozygous M_7
d — structurally homozygous M_7

it is very unlikely that genic activity in the secondary spermatocytes can be significant for the completion of meiosis. However, in the irradiation induced translocation heterozygotes of the mouse studied by LYON and MEREDITH (1966) there was a correlation between sterility and the configurations observed at diakinesis. Thus while uneven configurations (III + I) occurred with a frequency of 25.8% in sterile males their incidence in semisterile males was only 1.25%. The latter group was also characterised by a very high frequency of ring as opposed to chain multiples though the significance of this difference is not clear. It is true that MARTIN and HAYMAN

(1966) have suggested that chromosomal unbalance in the hare-wallaby *Lagorchestes conspicillatus* may be lethal. This was based on their failure to recover numerically unbalanced secondary spermatocytes in a male heterozygous for an autosomal fusion despite the fact that 30% of the first metaphase cells showed a linear trivalent. This is completely negative evidence and is certainly not borne out by the positive evidence available from F_1 hybrids in *Eyprepocnemis* (John and Lewis 1965) where unbalanced secondary spermatocytes appear to be perfectly functional. The same is true of polysomic grasshoppers (Fig. 81) and polysomic cells in *Pyrgomorpha* (Lewis and John 1959).

It would appear then that although the failure found in translocation (and trisomic) mice is not observed at a pre-meiotic stage it must, nevertheless, be regarded as genotypic in nature. In other words, it is the activity of the genotype which is disturbed. The question then is whether the events of breakage or reunion actually caused genic damage at the site of breakage, whether the rearrangement disturbed the organisation of some kind of super-genic operon so that the activity of essentially undamaged units was impaired (position effect) or whether the radiation which induced the rearrangements in the mouse also caused deletions at the site of exchange or distinct gene mutations at sites linked with the rearrangement. In view of the sterility, these possibilities cannot be resolved.

One thing can, however, be said. The viable rearrangements which have been studied in plants and animals are rarely associated with position effects. And, as we have seen, the Cattanach translocation which is known to have a position effect (see pg. 144) is fertile. White (1964), however, claims that "the great majority of newly arisen spontaneous chromosomal rearrangements probably have deleterious position effects associated with them." As "one proof" of this view White points to the fact that while there are no meiotic mechanical impediments to the establishment of structural rearrangements in many parthenogenetic species, such changes have not in fact been exploited by them. This argument is a curious one especially in view of the opinion which White next expresses, namely, that the significance of rearrangements is that they "permit the building up of coadapted blocks of genes which cannot be broken down by crossing-over." Clearly, if this latter view is correct then rearrangements can have no such role in non-meiotic parthenogenetic forms. What is more, the validity of White's claim regarding the absence of rearrangements in non-sexual types is questionable. Indeed White *et al.* (1963) have described a parthenogenetic species of grasshopper, *Moraba virgo*, with complex structural heterozygosity. Further, a vast range of cytotypes is often associated with apomixis in plants (see pg. 73). Whether this implies that rearrangements can have neutral, or even favourable, effects is not clear. But the claim that spontaneous rearrangements are normally associated with deleterious position effects is not easily justified. Indeed there seems good ground for concluding that provided all the chromosome segments are present their precise arrangement within the complement is of far less consequence. This is supported by the large number of structural rearrangements that persist and apparently cause at the most only trivial changes in morphology or physiology.

Four other examples implicate genic factors in karyotypic stabilisation. First, the stabilisation of the allotetraploid phase and the effective amphidiploidisation of both it and the hexaploid in the evolution of wheat has been perfected by what appears to be a simple mutation (Riley and Chapman 1961). Secondly, in *Narcissus bulbocodium* B-chromosomes are apparently normal members of the complement which became heterochromatic—and hence genetically isolated from their homologues—under

Fig. 82. The karyotype of *Oxalis dispar* (after Marks 1957).

Table 65. *Frequency of Telocentric (TE) Chromosomes per Cell in 30 Pollen Grains and 62 Seedlings of Oxalis dispar.*
(Data of Marks 1957.)

No. of TE chromosomes per grain		3		4		5
Pollen grains	No.	4		19		7
	%	13.3		63.3		23.3
Seedlings	%	6.5	35.5	43.5	12.9	1.6
	No.	4	22	27	8	1
No. of TE chromosomes per seedling		6	7	8	9	10

the action of a dominant gene H which exists in plants of certain wild populations (Fernandes 1949). Thirdly, Brewbaker and Natarajan (1960) have demonstrated the induction of self-fertility in a self-incompatible species as the result of the addition of supernumerary centric fragments carrying an incompatability (S) allele. They have argued that self-fertility of this type could confer considerable survival value on supernumerary fragments and so provide a selective mechanism for their establishment in a species.

Fourthly, *Oxalis dispar* has a complement of $2n = 12$ consisting of $2SM + 3ST + 7TE$ chromosomes (Fig. 82) and chromosome breakage at the centromere in the ST's may on occasion give rise to additional TE's. Pollen grain data (Table 65) show that cells with TE chromosomes are selected, for there is an excess of pollen grains with 4 and 5 telocentrics and a deficiency of those with 3. This depends on the selective survival of the TE segregants from the heteromorphic ST-TE bivalent which forms regularly at meiosis. Marks (1957) has given good grounds for believing that since the TE chromosomes lack the short arms present in ST's these arms may be deleterious. The seedling data (Table 65) certainly show the detrimental nature of the short arm of the ST chromosome of the heteromorphic pair. The number of seedlings with a homomorphic pair of ST chromosomes is much lower than expected from random segregation while

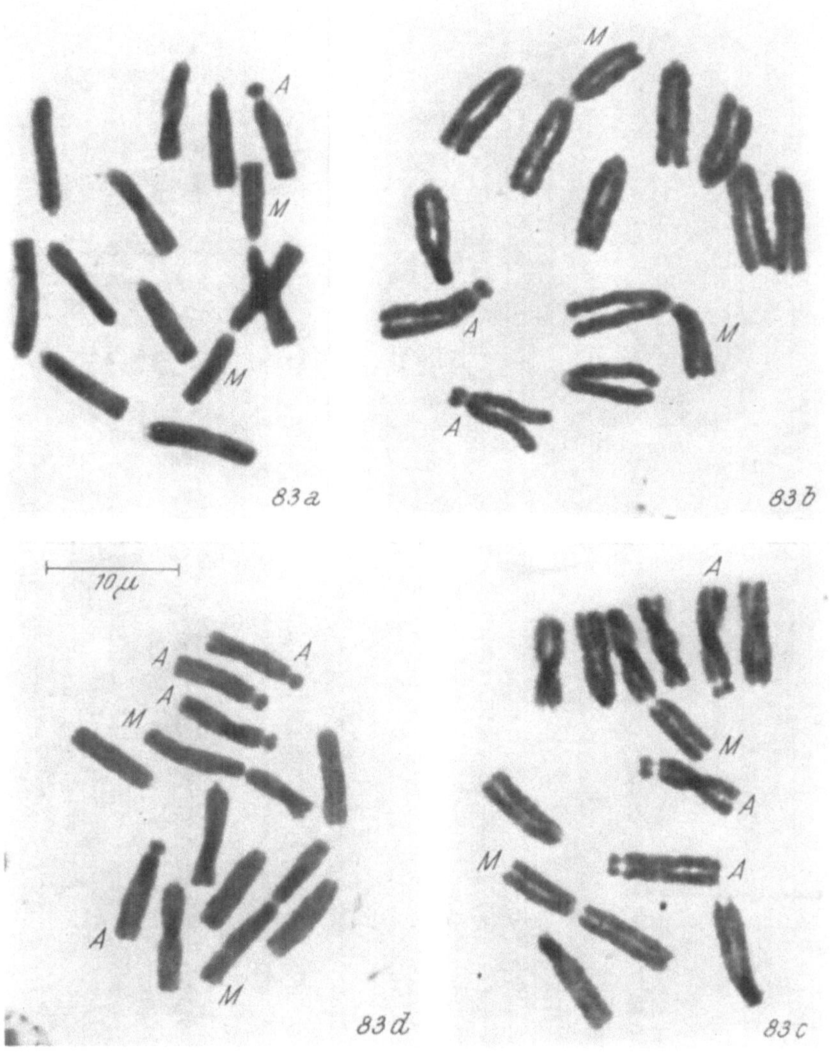

Fig. 83. Karyotype variation between individuals in the roots of *Oxalis* seedlings (compare with Fig. 82; these preparations were made from material kindly supplied by Dr. G. E. Marks).
a — one acrocentric (A)
b — two acrocentrics
c — three acrocentrics
d — four acrocentrics

that of seedlings with the homomorphic pair of TE chromosomes is greater (Fig. 83). Note however that chromosome behaviour and segregation on the female side is normal, and egg nuclei with 3 and 4 TE-chromosomes occur

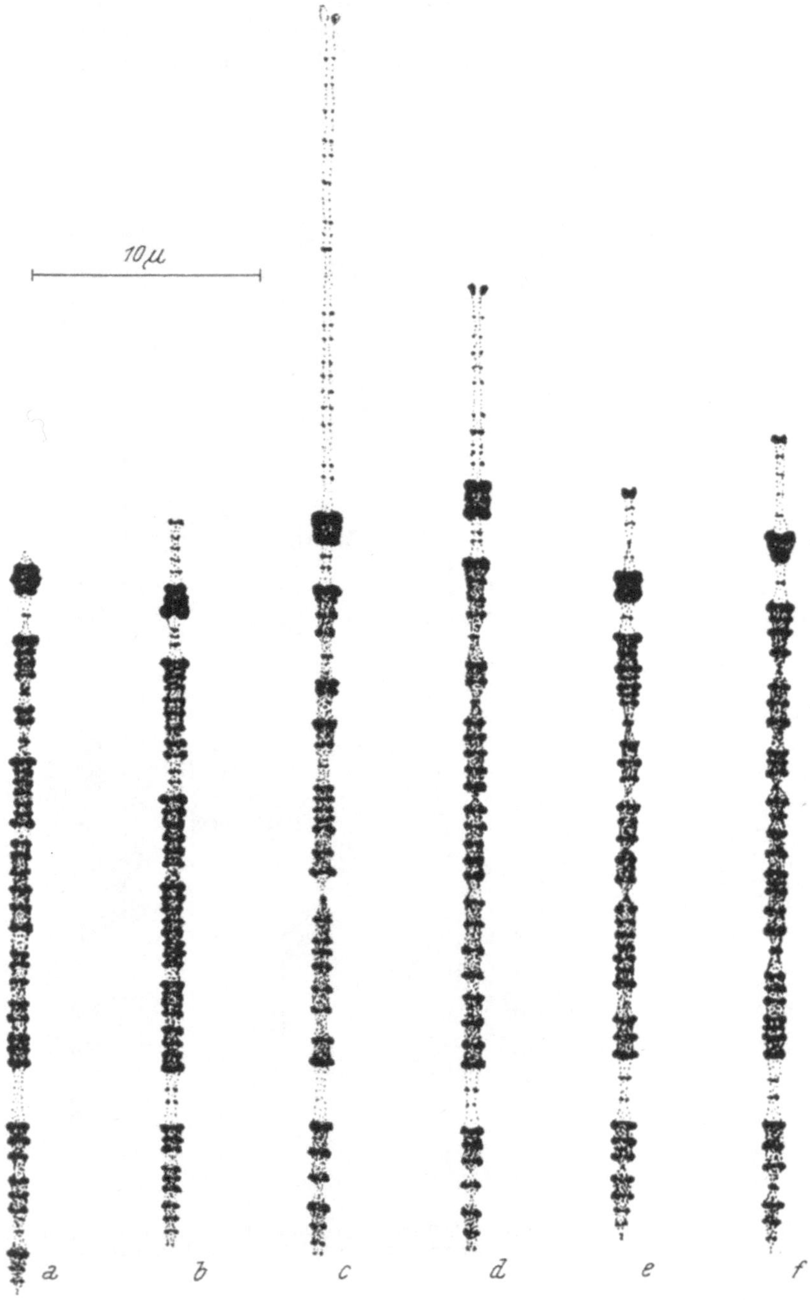

Fig. 84. Variation in the organisation of the terminal region of the standard B-chromosome in rye (after Lima-de-Faria 1963).

a — Sweden (Östgöta Grarag), cultivated
b — Sweden (Wasa), cultivated
c — Turkey, semi-wild
d — Afghanistan, semi-wild
e — Transbaikalia, cultivated
f — Korea, cultivated

with equal frequencies. Finally, since seedlings with 10 TE-chromosomes were found it would seem that misdivision of ST-chromosomes can occur rarely to give egg nuclei with 5 TE-chromosomes.

Of course, it is not always easy to distinguish between the mechanical and genetical facets of stabilisation; indeed both may interact to give a particular end product. For example, a comparison of the standard-type B-chromosome of rye from different geographical areas reveals that the same essential pattern is present in all the varieties examined. There are, however, differences in the size of the weakly stained distal region of the long arm (LIMA-DE-FARIA 1963). In Turkey this region is as long as the rest of the long arm and contains some 20 chromomeres. In Östgöta, on the other hand, this region is nearly absent (Fig. 84). The centre of origin of rye is considered to be in Asia Minor so that the B-type in Turkey and Afghanistan is probably the original one. Successive small deletions in the telomere region of the long arm have then led to the formation of the other B-types. The Wasa and Östgöta chromosomes have least stability and have given rise to several derivatives, the most extreme of which is the ring-B in Östgöta. It would appear, therefore, that the reduction of the telomere region may determine the higher instability of the Östgöta type.

In human leukaemias the X-6-12 chromosomes are often over-represented while the 13—15 and 21—22 chromosomes are under-represented. Likewise out of 40 human tumour stem-lines studied 27 had more chromosomes then expected in the X-6-12 group while 33 and 34 tumours respectively had fewer chromosomes than expected in the Y-21-22 and 13—15 groups (LEVAN 1966). Thus there are chromosome types which show a marked tendency to increase and others which tend to decrease.

As LEVAN points out it is suggestive that the tendency to decrease is shared by the two chromosome groups that have the smallest short arms, carry satellites which are involved in the formation of nucleoli and are known to be more unstable than any of the other members of the human complement. Thus the most common human trisomy is that of 21—22 and the most common translocations involve the short arms of the different members of the two groups. This non-randomness in advanced cancer stemlines can be accounted for in one of two ways. One could view the most common karyotypes as those which are best adapted for the life as cancer cells. Alternatively it is conceivable that it stems from the tendency of the 13—15 and 21—22 groups to undergo translocations which give rise to chromosomes which fall morphologically into the preferred X-6-12 group.

The standard complement of *Datura* consists of 12 pairs of metacentric chromosomes which are conventionally referred to as 1.2, 3.4, 5.6 ... 23.24 respectively. Seven of these chromosomes (3.4, 7.8, 9.10, 11.12, 19.20, 21.22 and 23.24) have a secondary constriction and in all but one of them (3.4), this takes the form of a satellite. In addition 13.14 has a compound constriction (see pg. 18). Of the seven chromosomes with secondary constrictions only one (9.10) has given rise to secondary trisomics involving the doubling of the arm with the satellite (10.10). In all other secondaries, doubling has occurred in the arm lacking the satellite (AVERY, SATINA, and

Rietsema 1959). A second feature of interest in this organism is that the presence of an extra chromosome in the parent increases the occurrence of unrelated (2 n + 1) types in the offspring.

The possible modes of interaction of stabilising factors is seen also in an experiment involving *Oenothera*. A haploid plant of *O. hookeri* arose spontaneously in 1936 and, following failure of reduction, homozygous diploids were produced. Linnert (1965) studied tetraploid derivatives obtained by the colchicine treatment of these diploids in 1941 and maintained by selfing for 15 generations. She also studied diploids derived from these tetraploids by the parthogenetic development of reduced eggs following pollination with irradiated pollen. The behaviour of these tetraploids and diploids was compared with that of normal diploids with regard to the incidence of chromosome mutations. These latter included translocations, trisomy and iso-chromosome aneuploidy. In 34 tetraploids a total of 31 different mutations were observed while in the normal diploid population only 3 aberrant individuals were found among the 100 tested. Clearly the tetraploid condition of the former favoured the maintenance of abnormal karyotypes. This was further borne out by the fact that the quota of mutants which were transmitted to the parthenogenetically derived diploids was lower and their frequency declined sharply and continuously during their sexual perpetuation.

Perhaps the clearest example of the interaction between mechanical and genetical factors is seen in relation to the meiotic behaviour of interchange complexes. Given random behaviour there is little hope for the stabilisation of interchange heterozygotes. Where therefore interchanges survive in the heterozygous form in nature their meiotic performance—and more especially their segregation—cannot be at random. The disjunction of interchange multiples depends upon both structural and genotypic aspects of chromosome behaviour. These include:

(i) Chiasma frequency—in rye the higher the chiasma frequency of interchange associations the lower the probability of disjunctional orientation (Rees and Sun 1965). Genotypes with higher chiasma frequencies thus tend to show lower disjunction.

(ii) Chiasma position—in chain multiples the position of an "interstitial" chiasma has a powerful if not unconditional influence on the orientation behaviour of the multiple (Lewis and John 1963). Indeed Rickards (1964) concluded from his study in *Allium triquetrum* that if interstitial chiasmata have formed then the number and positions of chiasmata in the other segments of interchange complexes are not important to segregation.

(iii) The stability of orientation patterns and their capacity for re-orientation—these properties depend of course on the distance between the centromeres, the relative lengths of the chromosome arms and the position of the chiasmata in the multiple system. Within the limits set by these properties the most frequent modes of orientation in multiple chromosomes associations are those with the highest stability (Bauer, Dietz, and Röbbelen 1961).

MOENS (1965) has drawn attention to a situation which may well influence the stabilisation of structurally modified chromosomes. Chromosome 2 in the tomato is satellited. Among the progeny of a plant trisomic for chromosome 2 was one which carried an iso-chromosome for the satellited short arm (Fig. 85). This arm does not carry any known genes and undergoes no crossingover. Consequently diploid plants with one additional iso 2s-2s are phenotypically normal and have normal fertility. From test crosses it was determined that transmission of this iso-element was 30% through both male and female gametes. More than two isochromosomes could not, however, be accumulated in any one plant.

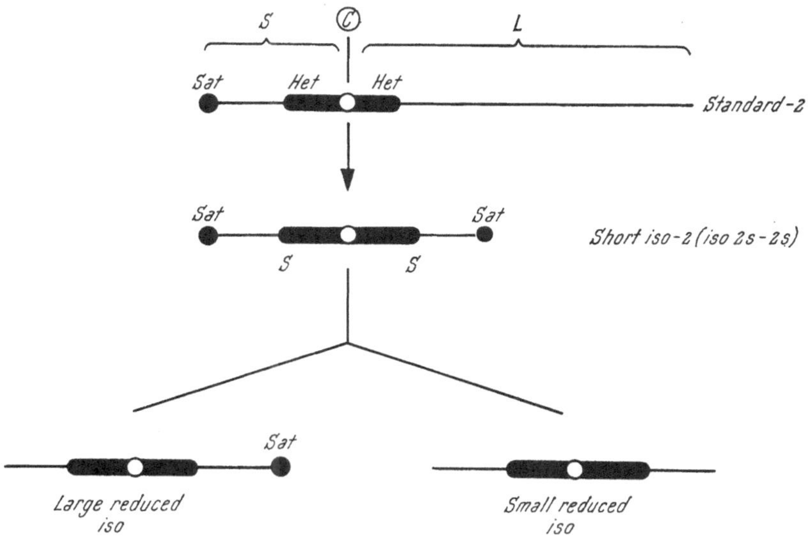

Fig. 85. Derivatives of chromosome-2 in tomato following misdivision (based on MOENS 1965).

Reduced iso's occurred regularly (Fig. 85). These were found to be less detrimental and could be accumulated to the extent of three per plant. In consequence the number of iso's per plant was found to increase for these reduced types. The moral of this case is that secondary changes can arise which are more favourable to the perpetuation of an extra chromosome. MOENS, therefore, suggests that extra chromosomes, or indeed other chromosomal abnormalities, which have a selective disadvantage at the time of their origin, but which are still present after several or many generations, may have acquired characteristics favourable to their perpetuation. Pertinent to this suggestion are some observations of HARLAN LEWIS in the plant genus *Clarkia*. Some species of *Clarkia*, but not *C. biloba*, are characterised by high frequencies of interchange heterozygotes in wild populations. These invariably show a regular alternate segregation. *C. lingulata* is also distinguished from *C. biloba* by a translocation which in the hybrid between the two species is not regular in its dissociation. *C. lingulata* differs further from *C. biloba* in at least two paracentric inversions. In an experiment designed to test the persistence of chromosome variation, populations were initiated in natural habitats by sowing seeds derived from hybrids between *biloba* and *lingulata*. Although this hybrid has very low fertility, a few seeds were obtained from extensive backcrossing. The backcross progenies, which included individuals of various fertilities and chromosome composition, were self-pollinated or backcrossed a second time to the parental species. Progenies from these crosses were then grown in a block on the experimental field and open pollinated seeds were used for sowing at three selected sites in several natural habitats in southern California.

Table 66. *Changes in the Chromosome Composition of a Synthetic Population Containing Hybrid Derivatives of Clarkia biloba* (n = 8) × *Clarkia lingulata* (n = 9). (Data of H. Lewis 1962.)

%-Frequency of Chromosome Type	Interchanges		L-Chromosome	Inversions	
	T_1	T_2		Plants with 1 Inversion	Plants with 2 Inversions
1. Initial (est.)	32	13	15	34	17
2. After 5 generations (n = 75)	43	14	4	29	14
% Difference	+9	+1	− 11	− 5	− 3

N. B. (1) T_1 = Translocation that distinguishes *C. lingulata* from *C. biloba*. This translocation greatly reduces fertility when heterozygous since it is non-disjuntional in behaviour.

(2) T_2 = Translocation from *C. biloba*. Heterozygotes for T_2 are disjunctional and it does not, therefore, cause any reduction in fertility.

(3) L-chromosome = Additional chromosome in *C. lingulata* (n = 9) that distinguishes it from *C. biloba* (n = 8).

(4) The two inversions cannot be distinguished cytologically.

From the data on fertility and cytology of the block of garden plants, estimates were made of the initial frequencies of the various chromosome arrangements in the samples sown. The chromosomal differences that were scored are summarised in Table 66. Of the three initial synthetic populations sown only one survived and 75 plants from it were examined after 5 generations. All the initial differences were still present. Neither of the two translocations had notably decreased in frequency and one of them —significantly the non-disjunctional one (T_1)—may even have increased. Chromosome differences producing low fertility as heterozygotes are thus not necessarily immediately eliminated but may persist in a population without significant change in frequency for several generations.

The persistence of these chromosomal differences in the experimental population for five generations suggests that new chromosome arrangements which arise spontaneously may remain for at least a few generations in a population without being eliminated even when the heterozygote has greatly reduced fertility. Henricson and Backstrom (1964) reported the case of a boar which while normal in respect of mating capacity and semen quality had a markedly reduced fertility. Thus from 51 sows served by this boar the average litter size was 5.1 piglets against a mean value of 12.7 piglets obtained from previous pregnancies of 21 of these 51 sows sired by different boars. This gives a fertility reduction of about 56%. On examination this boar proved to be an interchange heterozygote.

Sperlich (1966) has recently made a study of the fate of X-ray induced inversions in inbred lines of *Drosophila melanogaster*. Strains made isogenic for their 2nd and 3rd chromosomes were irradiated and nine inversions (including an overlapping double) were induced in chromosome III and four in chromosome II. Experimental populations were established carrying these induced inversions at initial gametic frequencies of 25%. In six of the populations the inversions were eliminated, or were close to elimination, within a year (Table 67). Three populations were lost by contamination and in the two remaining populations the inversion systems established what Sperlich claimed to be balanced equilibria. This, if correct, is all the more remarkable because these two inversions proved to be lethal, or at least sublethal, in homokaryotypes.

Table 67. *Frequency Changes in Nine X-ray Induced Inversions Introduced Into Experimental Populations of Drosophila melanogaster at an Initial Gametic Frequency of 25%.* (Data of Sperlich 1966.)

Inversion No.		July 64 (100 Chrms.)	Aug. 64	Sept. 64	Dec. 64	May 65	Fate of Inversion
			(300 Chromosomes per Pop.)				
Chromosome III	1	9.0	9.7	9.0	6.0	—	Eliminated
	2	32.0	10.0	3.7	0.0	—	
	3	23.0	16.7	19.7	2.0	—	
	4	19.0	15.0	14.7	24.7	5.0	
	5	20.0	13.3	13.3	8.0	0.3	
	6	18.0	11.3	11.0	—	—	
	7	16.0	20.7	20.0	29.3	19.0	Balanced polymorphism
Chromosome II	8	7.0	12.3	10.7	14.7	14.0	
	9	13.0	18.3	10.0	4.0	2.0	Eliminated

One final question concerns the relationship between changes in the germ line as opposed to the soma. As we have seen (pg. 36) a series of precisely controlled eliminations take place in the course of the normal life cycle of *Sciara*. Those which occur in the germ line differ from those in the soma and that in the male soma differs from its counterpart in the female. The available evidence indicates that while the soma cannot sustain extensive losses and/or additions of genetic material the germ line can. Crouse (1965), for example, has utilised X-ray induced interchanges between X and IV (OR T 26 and OR T 27) in *S. coprophila* to produce aneuploids which are both trisomic for a region of IV and monosomic for a region of X. In the case of OR T 26 the aneuploids develop into fertile females but in OR T 27 they die either as late 4th instars or else as pupae. Their gonads, however, develop and show typical ovarian differentiation. Aneuploid males are not recovered among the progeny of aneuploid females and clearly the male soma cannot tolerate the aneuploid state. Males with euploid soma and aneuploid germ line can, however, be produced and these are both viable and fertile. Thus, although the male soma cannot tolerate any one of the aneuploid conditions the germ line develops normally.

3. The Integration of Karyotypic Changes

Once novel karyotypes become stabilised their integration may follow a variety of patterns.

Ehrendorfer (1960) has analysed the factors involved in the integration of B-chromosomes in *Achillea asplenifolia* by introducing B's into an experimental hybrid population and then analysing 300 individuals in terms of their B-content. This study indicates a direct capacity for autoregulation towards an optimum, and more or less stable, dosage of two. Direct autoregulation is achieved mainly by intra-individual somatic reduction in plants with higher numbers and multiplication by non-disjunction in premeiotic mitosis in plants with only 1 B. Indirect regulation of the number of B's depends upon the fact that the dosage of B-chromosomes influences compatibility and pollen tube competition as well as fertility. Direct and indirect regulation in both experimental and natural populations thus leads to a self perpetuating polymorphism of the balanced type (Fig. 86). This example illustrates admirably the complexity of the problem of integration and the multiplicity of interacting factors which may be involved in the process of integration.

Kimura (1962) has argued that any supernumerary must be endowed with a meiotic-drive-mechanism from the outset if it is to be permanently maintained in a population. Such a system of meiotic drive leads to segregation bias in the transmission mechanism and Kimura and Kayano (1961) have shown how, in theory, relatively large but inert supernumeraries can evolve as a result of segregation distortion protecting them from loss by selection. Kimura's argument is, however, based on the assumption that a supernumerary, which must initially be derived from a standard member of the chromosome set, creates a trisomic state which must be deleterious and can survive only if it possesses a meiotic system sufficiently strong to counterbalance its selective disadvantage. This condition certainly appears to exist in the mealy bug (Nur 1966). We do well to remember, however, that while fertility is very important it is but one aspect of fitness.

As far as selective integration is concerned different evolutionary strategies result according to the nature of the chromosome system concerned. Thus in the case of structural variants the mutant state may be:

(a) fixed in the homozygous condition following the secondary formation of structural homozygotes,

(b) preserved and integrated in a state of permanent heterozygosity, or else,

(c) a balance may be created between homozygotes and heterozygotes—either one type of heterozygote and one type of homozygote, as in most sex-chromosome systems (see pg. 92), or more than one type of each as is usual in the case of autosomal polymorphism (see pg. 73).

Both structural and numerical changes, of any kind, have a better chance of reaching fixation or developing into a polymorphic facies when they arise in communities of restricted size. Thus, to be identical, structural

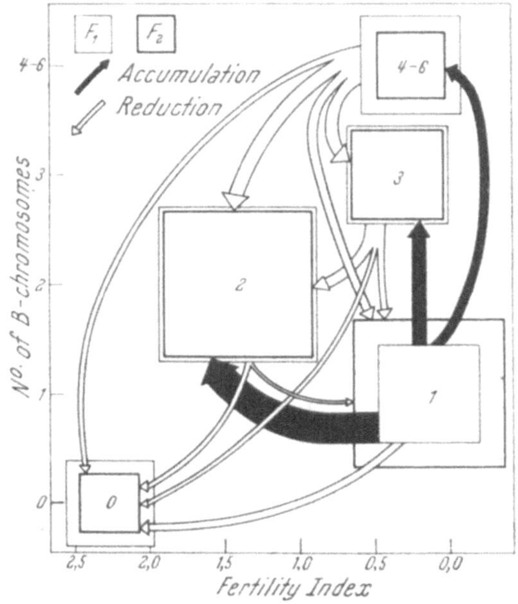

Fig. 86. Factors influencing the autoregulation of B-chromosome frequency in two generations (F_1 and F_2) of an experimental hybrid population of *Achillea asplenifolia* and *A. setacea*. The size of the squares gives a measure of the number of individuals at a given B-chromosome level. The thickness and direction of the arrows indicates the intensity and direction of the two principal regulatory mechanisms, somatic reduction (open arrows) and pre-meiotic accumulation (solid arrows). Balance is maintained at the optimum level of two B-chromosomes per individual (after EHRENDORFER 1960). The fertility index is based on the number of seedlings per capitulum.

changes must have both sets of exchange points in precisely the same positions. This, instead of merely doubling the rarity of a given type of 2-break rearrangement raises it to the square. In turn this implies that, in any given population, all identical chromosome mutants must be descended from a single original nucleus in which that unique rearrangement first arose. Structural changes are, thus, invariably monophyletic and this must considerably hinder their establishment. Under what conditions then can the mutant become integrated into the chromosome system of the species? The most obvious condition is that of reduced competition and selection pressure. This, in turn, may be expected to arise following a change in the environment which is detrimental to the wild type or else from a change of environment following migration.

The reduction in size of the population which must follow in these situations will impose a change toward greater inbreeding and thus to greater homozygosity. This, if unopposed, must often lead to genotypic unbalance. Chromosome behaviour is genotypically controlled and genotypic unbalance can lead to increased chromosome mutation. What more can one ask for?—Mutation, reduced competition and a change in the direction of selection.

From the point of view of selective criteria there is good ground for arguing that changes in the chromosome complement are concerned with restrictions to, or extensions of, the recombination system (JOHN and LEWIS 1966). Inversions, for example, when heterozygous, may be equally potent conservators of the segments which they span as they are effective boosters of recombination elsewhere in the chromosome complement. They therefore offer a considerably more flexible method of regulating recombination than simple chiasma localisation. It is true that many structural control systems have less capacity for reversibility than systems of direct genotypic control. But structural changes can preserve or fix genotypes with far greater effectiveness than is possible through genotypic regulation of recombination.

Allopolyploidy too can be viewed as a means of stabilising and integrating recombinant or intermediate genotypes. In *Dactylis*, for example, diploid hybrids are fully fertile and vigorous in the F_1 and later generations. The adoption of polyploidy by such hybrids cannot then have the function of restoring fertility. It does, however, tend to stabilise a condition intermediate between the parental types because of the complexities of tetrasomic segregation and the larger number of intermediate conditions which this type of segregation makes possible (STEBBINS 1966). Allopolyploidy, then, must often have the adaptive function of stabilising genotypes which are intermediate between, or represent recombinations of, the characters found in ancestral diploids. Indeed ZOHARY and NUR (1959) have shown how the same process can also operate in autopolyploid systems. The polytypic group of the orchard grass (*Dactylis glomerata*) has a large, sexual, tetraploid superstructure ($2n = 28$). Cytogenetically all tetraploid forms are interfertile. Natural triploids also occur at a low frequency. These are vigorous and normal in appearance and cannot be distinguished morphologically from the adjacent diploids and tetraploids. Seeds collected from such triploids when planted under greenhouse conditions give rise to tetraploid (12 plants) and pentaploid (6 plants) progeny as well as aneuploids (4 plants). These can be accounted for in terms of the production of unreduced eggs in the triploid (Fig. 87) and admirably illustrates the artificiality of the distinction between auto and allo-polyploids.

The evolution of polyploids illustrates also the temporal aspect of the problem of integration. Studies on newly produced, artificial, polyploids have shown that levels above tetraploidy are rarely tolerated when they arise directly from diploid forms (GREENLEAF 1938). Successful polyploids may, however, become diploidised over long periods of time so that further elevation of number is possible by successive cycles of ploidy. Given a sufficiently long span of time, accompanied by diploidisation at successively higher chromosomal levels, almost no limit appears to exist to the upward trend of polyploidy (STEBBINS 1966).

One of the principal genetic requirements during integration has been the necessity to produce temporary constancy through the promotion of linkage. In plants this has involved the integration of three types of karyotypic change—reduction of basic number, change from median to subterminal kinetochore position and a trend toward increasing size difference between members of the same complement (STEBBINS 1958). The net result of this is to convert symmetrical karyotypes (chromosomes all metacentric and about the same size) into asymmetrical ones (many chromosomes with subterminal centromeres and with a wide range in size). Thus in the *Cichorieae*, STEBBINS has argued that the original members of the

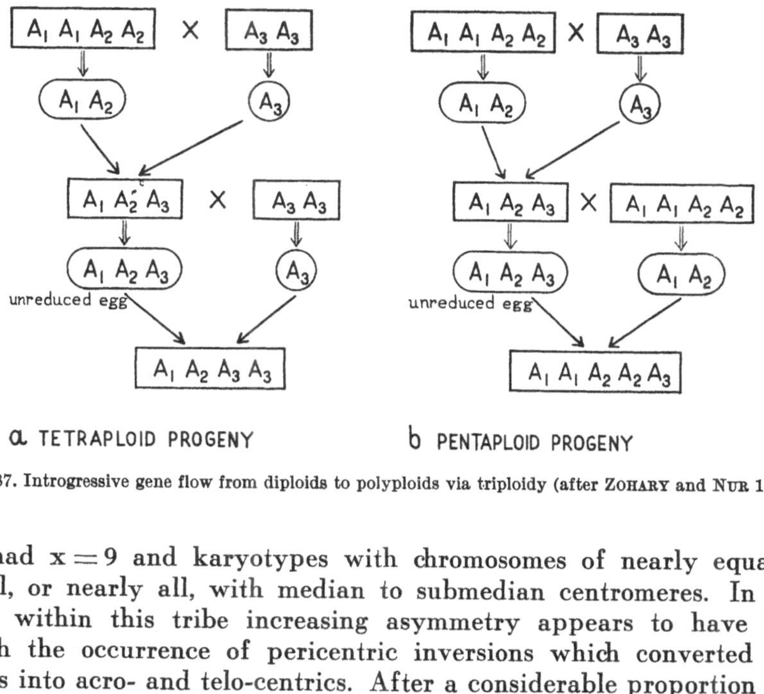

Fig. 87. Introgressive gene flow from diploids to polyploids via triploidy (after ZOHARY and NUR 1959).

tribe had $x = 9$ and karyotypes with chromosomes of nearly equal size and all, or nearly all, with median to submedian centromeres. In many genera within this tribe increasing asymmetry appears to have begun through the occurrence of pericentric inversions which converted metacentrics into acro- and telo-centrics. After a considerable proportion of the chromosomes had acquired subterminal centromeres a second type of asymmetry developed through the occurrence and establishment of unequal translocations between non-homologous chromosomes. The principal effect of this change was to accentuate size differences between chromosomes. Asymmetry in size thus followed asymmetry in centromere position and size asymmetry actually led to a decrease in centromere asymmetry. The net result of these changes is to produce small chromosomes as reciprocal products. These are likely to be heterochromatic since heterochromatin is located proximally in the chromosomes of this tribe. This, in turn, is followed finally by the loss of some of these small heterochromatic remnants and hence a reduction in number. This sequence of integration appears to be paralleled in *Aster* (*Compositae*), *Aegilops* (*Hordeae—Gramineae*), *Aconitum* and *Delphinium* (*Helleboreae—Ranunculaceae*).

The operation of such a system of integration, whereby similar changes have occurred and established themselves repeatedly in an evolutionary lineage, WHITE (1964) has termed karyotypic ortho selection. Both occurrence and establishment depend, as we have seen, on the mechanical and genetical properties of the five organellar components which we enumerated at the beginning of this monograph. It is in terms of the deployment and redeployment of these organelles that the chromosome complement can be defined and refined.

For many purposes we can consider the chromosome system as an entity divorced from the reproductive and other life processes of a species. But when we come to consider integration it is necessary to relate the chromosome and the breeding processes of a species. For example, there are two general ways in which a sterile plant hybrid can undergo chromosome doubling. First by somatic doubling in the hybrid. Second by the production and union of unreduced gametes in diploid sterile flowers. Somatic doubling has a greater chance of occurring in a long-lived perennial plant than in an annual. Thus the first tetraploid branch of *Primula kewensis* did not develop during the first year of the life of the hybrid plant. If, therefore, the parental species had been an annual, tetraploid *Primula kewensis* might never have originated (GRANT 1965). It is not surprising therefore that polyploid species are known to occur with a higher frequency among herbaceous perennials than among annuals. Of course, even where the time available for somatic doubling is reduced it is still possible to get unreduced gametes. The frequency of production of unreduced gametes will depend on the nature of meiosis in the diploid hybrid. Thus the spores produced by the diploid *kewensis,* where pairing is high, were almost exclusively reduced with only 9 chromosomes. Those of the diploid *Rhaphanobrassica* hybrid, on the other hand, where there is virtually no pairing, included a large number which were unreduced. Significantly the tetraploid form of this hybrid arose from seed. The subsequent perpetuation of new forms like these is clearly facilitated by a closed breeding system. It is presumably for just such a reason that polyploid species are far more frequent among autogamous annuals than in cross-fertilising annuals.

The length, shape, number and structure of the chromosomes which make up the complement are thus not at random. Each species is characterised by a remarkably strict organisation of its chromosome complement, an organisation which defines its karyotype. The karyotype is characterised by, and can be defined in terms of, a stable number of chromosomes each with its own stable structure. For the sequences of heterochromatic segments, nucleolar organisers and centromeres determine not only the mechanical but also the genetical characteristics of a species. The genotype created by evolution can only be efficiently and successfully transmitted in a particular chromosome guise. It is true that in some groups the structure of the chromosomes is variable while their numbers are constant. In others a variety of chromosome numbers may be found predicated on a uniformity of structure. But in all cases the types that we see are but a small and selected sample of the enormous range of variants which are developed during the life of an organism. The greater part of this variation is rejected in each generation. The extent of this rejection is especially severe in cases where the sexual generation is lengthened. Thus among slow-growing long-lived shrubs and trees there is a remarkable constancy of chromosome number. Likewise B-chromosomes do not appear in woody plants (Darlington 1956).

Résumé

The chromosomes are the vehicles of mendelian heredity and, as such, are the principal DNA-containing structures of the cell. The function of heredity is, of course, development so that development—or, rather, differential development—reveals heredity. Thus, not only are the chromosomes the principal basis of the genetic system but of the epigenetic system as well. This means, amongst other things, that the chromosomes have both a genotypic and a phenotypic aspect. The genetic variations within and between chromosomes and the mutations on which they rest are, however, revealed only fractionally at the level of chromosome morphology. And the extent of this revelation varies both with the chromosome and the stage at which it is studied. Indeed it may even be concealed as in the case of isomorphic chromosomes (Manna and Smith 1959, see pg. 91). These represent a heterogeneous population of chromosomes with regard to genetic structure and function, yet masquerading under the same morphological guise (see Fig. 48).

Likewise the differential behaviour which underlies their phenotypic variation is expressed at the cytological level to only a limited extent. Thus, in respect of both types of variation, the common character of chromosomes tends to conceal some of their uniqueness. This imposes a limitation on the technique of looking at chromosomes. Even so they remain the most distinctive and individual of all cell structures and they reward research more objectively than any other cell organelle.

In general, genetic changes at the chromosome (versus gene) level are important in relation to the endophenotype and evolution (John and Lewis 1966). Consequently, they are seen for the most part between individuals and, more especially, between breeding groups. Phenotypic changes in the chromosomes, on the other hand, are concerned with development and differentiation and they occur within individuals. These general expectations with regard to chromosome variation, and the parallel ones in relation to constancy, admit of exceptions however. Thus the chromosomes are not always constant within a breeding group, neither do they always remain genetically unchanged during development.

Now, it is quite clear that even the complex aneuploid mosaicism which characterises development in *Sciara* cannot determine cell differentiation though it does appear to be concerned in sexual dimorphism. But it is equally clear that variation of this kind is far from incidental to development. This can be inferred both from the complicated nature of the events concerned and the precision with which they are controlled. It is demonstrated too by the failure of the terminal stages of germ line development when the sequence of mosaic development is disturbed experimentally.

Such changes cannot, of course, be regarded as explanations of differentiation. They are, however, symptoms of it which, like any others, require an explanation. Indeed their study could reveal the mechanisms by which the chromosomes control themselves. In fact investigations in this direction have already exposed a variety of control systems, some canalised and confined within chromosomes (*Sciara*) or even parts of them (V-type position effects) and others which are operative between chromosomes (neo-centric activity, B-chromosome effects). So far, however, while some of the properties of the various communication channels have been described, the nature of the communications and their reception have not yet been elucidated (Lewis and John 1963, John and Lewis 1965).

There is, however, some information on the nature of the control in the other direction. Thus just as various inducers and repressors, internal or external, are known to act at the level of gene transcription, so too are substances like ecdysone, and their effect can be seen at the chromosome level in salivary glands. The study of development at the chromosome level is, however, a comparatively recent trend and by far the majority of karyological studies have been concerned with the chromosomes of the germ line, especially at meiosis. Variation in the chromosome complement at this level is not seen within individuals except as a consequence of meiotic segregation. And even some of this is effectively removed by complementary gametic elimination.

The nature of the germ line karyotype is important in two main directions. It is concerned directly with the control of recombination within cells and indirectly with the control of mating between individuals. The one leads to the other and to the definition of breeding groups. The course of evolution and its mechanism are thus brought together and the significance of chromosome change in the origin of species underlies and underlines its importance in the classification of those species. Thus, the chromosome complement is central to the life and reproduction of cells. It is central also to the species and its evolution. And its study leads one in both directions. For the chromosome represents a higher level of organisation and integration than that implied in a mere string of genes. And, though they are related, the principles governing its change and evolution are different from those of the genotype it carries, the exophenotype it determines and the endophenotypic system by which it is transmitted.

References

Alonso, P., and J. Perez-Silva, 1965: Giant chromosomes in Protozoa. Nature (London) **205**, 313—314.

Ammermann, D. von, 1965: Cytologische und genetische Untersuchungen an den Ciliaten *Stylonchia mylitus* Ehrenberg. Arch. Protistenk. **108**, 109—152.

Anderson, N. G., 1966: Cell division I. A theoretical approach to the primeval mechanism, the initiation of cell division and chromosomal condensation. Quart. Rev. Biol. **31**, 169—199.

Austin, C. R., 1960: Anomalies of fertilisation leading to triploidy. J. Cell. Comp. Physiol. **56** (Suppl. 1), 1—15.

Aula, P., 1965: Virus-associated chromosome breakage. A cytogenetic study of chickenpox, measles and mumps patients and of cell cultures infected with measles virus. Ann. Acad. Sci. Fenn. A IV Biol. **88**, 1—78.

Avery, A. G., S. Satina, and J. Rietsema, 1959: Blakeslee: The genus *Datura*. Ronald Press, New York.

Bajer, A., 1959: Change of length and volume of mitotic chromosomes in living cells. Hereditas (Lund) **45**, 579—596.
— 1965: Subchromatid structure of chromosomes in the living state. Chromosoma (Berlin) **17**, 291—302.
Baker, W. K., 1953: V-type position effects of a gene in *Drosophila virilis* normally located in heterochromatin. Genetics **38**, 328—344.
Bantock, C., 1961: Chromosome elimination in *Cecidomyidae*. Nature (London) **190**, 466—467.
Barigozzi, C., 1946: Über die geographische Verbreitung der Mutanten von *Artemia salina* Leach. Arch. Klaus-Stift. Vererb. Forsch. **21**, 479—482.
Barr, M. L., D. H. Carr, A. Morishima, and M. M. Grumbach, 1962: An XY/XXXY sex chromosome mosaicism in a mentally defective male patient. J. Mental Deficiency Res. **6**, 65—74.
Basrur, V. R., 1967: Inversion polymorphism in the midge *Glyptotendipes barpipes* (Staeger). Chromosoma (Berlin) **8**, 597—608.
— and K. H. Rothfels, 1959: Triploidy in natural populations of the black fly *Cnephia mutata* (Malloch). Canad. J. Zool. **37**, 571—589.
Battaglia, E., 1964 a: Cytogenetics of B-chromosomes. Caryologia (Firenzia) **17**, 245—299.
— 1964 b: Una mutazione con B-chromosomi (2 n = 14 + 3 B) *in Scilla autumnalis* L. (*Liliaceae*). Caryologia (Firenzia) **16**, 609—618.
Bauer, H., 1938: Die Polyploidie-Natur der Riesenchromosomen. Naturwiss. **26**, 77—78.
— 1947: Karyologische Notizen I. Über generative Polyploidie bei Dermapteren. Z. Naturforschg. **2**, 63—66.
— und W. Beermann, 1952: Der Chromosomencyclus der Orthocladiinen (*Nematocera, Diptera*). Z. Naturforschg. **7 b**, 557—563.
— R. Dietz und Ch. Röbbelen, 1961: Die Spermatocytenteilungen der Tipuliden III. Das Bewegungsverhalten der Chromosomen in Translokationsheterozygoten von *Tipula oleracea*. Chromosoma (Berlin) **12**, 116—189.
Bayreuther, K., 1954, Die Chromosomen der Flöhe (*Aphaniptera*). Naturwiss. **41**, 309.
— 1960: Chromosomes in primary neoplastic growth. Nature (London) **186**, 6—9.
Beatty, R. A., 1967: Parthenogenesis and polyploidy in mammalian development. University Press, Cambridge.
Beaudry, J. R., 1963: Studies on *Solidago* L. VI. Additional chromosome numbers of taxa of the genus *Solidago*. Canad. J. Genet. Cytol. **5**, 150—174.
Beçak, W., M. L. Beçak, and H. R. S. Nazareth 1962: Karyotypic studies of two species of South American snakes (*Boa constrictor amarali* and *Bothrops jararaca*). Cytogenetics **1**, 305—313.
Beçak, M. L., W. Beçak, and M. N. Rabello, 1966: Cytological evidence of constant tetraploidy in the bisexual South American frog *Odontophrymis americanus*. Chromosoma (Berlin) **19**, 188—193.
— — — 1967: Further studies on polyploid amphibians (*Ceratophrydidae*) I. Mitotic and meiotic aspects. Chromosoma (Berlin) **22**, 192—201.
Beermann, W., 1952: Chromomerenkonstanz und spezifische Modifikation der Chromosomenstruktur in der Entwicklung und Organdifferenzierung von *Chironomus tentans*. Chromosoma (Berlin) **5**, 39—108.
— 1954: Weibliche Heterogametie bei Copepoden. Chromosoma (Berlin) **6**, 381—396.
— 1962: Riesenchromosomen. Protoplasmatologia **VI**, D. Springer-Verlag, Wien.
Beermann, S., 1959: Chromatin-Diminution bei Copepoden. Chromosoma (Berlin) **10**, 504—514.
— 1966: A quantitative study of chromatin diminution in embryonic mitoses of *Cyclops furcifer*. Genetics **54**, 567—576.
Benazzi, M., 1966: Considerations on the neoblasts of planarians on the basis of certain karyological evidence. Chromosoma (Berlin) **19**, 14—27.
Bender, M. A., 1964: Chromosome breakage in *vitro*. Pp. 87—107 in "Mammalian cytogenetics and related problems in radiobiology". Pergamon Press, London.
Bennett, M. D., and H. Rees, 1967: Natural and induced changes in chromosome size and mass in meristems. Nature (London) **215**, 93—94.
Benirschke, K., and L. E. Brownhill, 1962: Further observations on marrow chimerism in marmosets. Cytogenetics **1**, 245—257.

Benirschke, K., and L. E. Brownhill, 1963: Heterosexual cells in testes of chimeric marmoset monkeys. Cytogenetics 2, 331—341.

Benyesh-Melnick, M., H. F. Stich, F. Rapp, and T. C. Hsu, 1964: Viruses and mammalian chromosomes. III. Effect of Herpes zoster virus on human embryonal lung cultures. Proc. Exp. Biol. Med. 117, 546—549.

Berendes, H. D., F. M. A. van Breugel, and Th. K. H. Holt, 1965: Experimental puffs in salivary gland chromosomes of *Drosophila hydei*. Chromosoma (Berlin) 16, 35—46.

Berger, C. A., and E. R. Witkus, 1954: The cytology of *Xanthisma texanum* D. C. I. Differences in the chromosome number of root and shoot. Bull. Torrey Bot. Chib 81, 489—491.

Berrie, G. K., 1959: Mitosis in the microspore of *Encephalartos berteri* Carruth. Nature (London) 183, 834—835.

— 1960: The chromosome numbers of liverworts (Hepaticeae and Anthocerotae). Trans. Brit. Bryol. Soc. 3, 688—705.

Bhattacharya, S. S., 1958: Die Wirkung von Röntgenstrahlen auf Kerne mit verschiedener heterochromatischer Konstitution. Chromosoma (Berlin) 9, 305—318.

Bier, K., 1957: Endomitose und Polytänie in den Nährzellkernen von *Calliphora erythrocephala* Meigen. Chromosoma (Berlin) 8, 493—522.

Blaser, H. W., and J. Einset, 1948: Leaf development in six periclinal chromosomal chimeras of apple varieties. Amer. J. Bot. 35, 473—482.

Böök, J. A., and B. Santesson, 1960: Malformation Syndrome in man associated with triploidy. Lancet i, 858—859.

Boothroyd, E. R., 1953: The reaction of *Trillium* pollen tube chromosomes to cold treatment during mitosis. J. Heredity 44, 3—9.

Boring, A. M., 1913: The chromosomes of the *Cercopidae*. Biol. Bull. (Woods Hole) 24, 133—146.

Bosemark, N. O., 1954: On accessory chromosomes in *Festuca pratensis* II. Inheritance of the standard type of accessory chromosomes. Hereditas (Lund) 40, 425—437.

— 1957: Further studies on accessory chromosomes in grasses. Hereditas (Lund) 43, 236—297.

Boyes, J. W., 1952: A multiple sex-chromosome mechanism in a root maggot *Hylemyia fugax*. J. Heredity 43, 195—199.

Bray, P. F., and A. Jospehine, 1963. An XXXYY Sex-chromosome anomaly. J. Amer. Med. Assoc. 183, 179—182.

Brewbaker, J. L., and A. T. Natarajan, 1960: Centric fragments and pollen-part mutation of incompatibility alleles in *Petunia*. Genetics 45, 699—704.

Brink, R. A., and D. C. Cooper, 1944: The antipodals in abnormal endosperm behaviour in *Hordeum jubatum* × *Secale cereale* hybrid seeds. Genetics 29, 391—406.

Brown, D. D., and J. B. Gurdon, 1964: Absence of ribosomal RNA synthesis in the anucleolate mutant of *Xenopus laevis*. Proc. Natl. Acad. Sci. (Wash.) 51, 139—146.

Brown, S. W., 1949: The structure and meiotic behaviour of the differentiated chromosomes of tomato. Genetics 34, 437—461.

— 1960: Spontaneous chromosome fragmentation in the armoured scale insects (*Coccoïdea Diaspididae*). J. Morph. 106, 159—185.

— 1963: The *Comstockiella* system of chromosome behaviour in the armoured scale insects (*Coccoïdea: Diaspididae*). Chromosoma (Berlin) 14, 360—406.

— 1966: Heterochromatin. Science 151, 417—425.

— and U. Nur, 1964: Heterochromatic chromosomes in the coccids. Science 145, 130—136.

Bush, G. L., 1962: The cytotaxonomy of the larvae of some Mexican fruit flies in the genus *Anastrepha* (*Tephritidae, Diptera*). Psyche 69, 87—101.

— 1966: Female heterogamety in the family *Tephritidae* (*Acalyptratae, Diptera*). Amer. Nat. 100, 119—126.

Caldecot, R. S., and L. Smith, 1952: A study of X-ray induced chromosomal aberrations in barley. Cytologia (Tokyo) 17, 224—242.

Callan, H. G., 1942: Heterochromatin in *Triton*. Proc. Roy. Soc. London B 130, 324—335.

— and L. Lloyd, 1960: Lampbrush chromosomes of crested newts *Triturus cristatus* (Laurenti). Phil. Trans. Roy. Soc. London B 243, 135—219.

CAMENZIND, R., 1966: Die Zytologie der bisexuellen und parthenogenetischen Fortpflanzung von *Heteropeza pygmaea* Winnertz, einer Gallmücke mit padogenetischer Vermehrung. Chromosoma (Berlin) **18**, 123—152.

CARR, D. H., M. L. BARR, and E. R. PLUNKETT, 1961: An XXXX sex chromosome complex in two mentally defective females. Canad. Med. Assoc. J. **84**, 131—137.

CARROL, M., 1920: An extra dyad and an extra tetrad in the spermatogenesis of *Camnula pellucida* (Orthoptera): numerical variation in the chromosome complex within the individual. J. Morph. **34**, 375—455.

CATTANACH, B. M., 1961: A chemically-induced variegated-type position effect in the mouse. Z. Vererbungslehre **92**, 165—182.

— 1964: Autosomal trisomy in the mouse. Cytogenetics **3**, 159—166.

CELARIER, R. P., 1955: Cytology of the Tradescantiae. Bull. Torrey Bot. Club **82**, 30—38.

CHEN, T. R., and A. W. EBELING, 1966: Probable male heterogamety in the deep sea fish *Bathylagus wesethi* (Teleostei: Bathylagidae). Chromosoma (Berlin) **18**, 88—96.

CHRISTENSEN, B., 1961: Studies on cyto-taxonomy and reproduction in the Enchytraeidae. Hereditas (Lund) **47**, 387—450.

— 1966: Cytophotometric studies on the DNA content in diploid and polyploid Enchytraeidae (Oligochaeta). Chromosoma (Berlin) **18**, 305—315.

— and J. JENNSEN, 1964: Subamphimictic reproduction in a polyploid cytotype of *Enchytraeus lacteus* Nielsen and Christensen (Oligochaeta, Enchytraeidae). Hereditas (Lund) **52**, 106—118.

CHU, E. H. Y., 1961: Chromosomal stabilisation of cell strains. National Cancer Institute Monograph No. 7, 55—71.

— 1964: Abnormalities of sex chromosomes. Pp. 209—232 in "Mammalian cytogenetics and related problems in radiobiology". Pergamon Press, London.

— and N. H. GILES, 1958: Comparative chromosomal studies on mammalian cells in culture I. The HeLa strain and its mutant clonal derivatives. J. Natl. Cancer Inst. **20**, 383—401.

— H. C. THULINE, and D. E. NORBY, 1964: Triploid-diploid chimerism in a male tortoiseshell cat. Cytogenetics **3**, 1—18.

CLAUSEN, J., 1951: Stages in the evolution of plant species. Cornell Univ. Press, Ithaca, New York.

CLEVER, U., 1961: Genaktivitäten in den Riesenchromosomen von *Chironomus tentans* und ihre Beziehungen zur Entwicklung I. Genaktivierungen durch Ecdyson. Chromosoma (Berlin) **12**, 607—675.

— 1963: Puffing changes in incubated and ecdysone treated *Chironomus tentans* salivary glands. Chromosoma (Berlin) **17**, 309—322.

— 1966: Chromosomal changes associated with differentiation. Pp. 242—252 in "Genetic control of differentiation." Brookhaven Symp. Biol. No. **18**.

CLOWES, F. A. L., 1961: Apical meristems. Blackwells, Oxford.

COHEN, M. M., and H. F. CLARK, 1967: The somatic chromosomes of five crocodilian species. Cytogenetics **6**, 193—203.

— and L. PINSKY, 1966: Autosomal polymorphism via a translocation in the guinea pig, *Cavia porcellus* L. Cytogenetics **5**, 120—132.

COLE, C. J., and C. H. LOWE, 1967: Sex-chromosomes in lizards. Science **155**, 1028—1029.

COOPER, K. W., 1951: Compound sex chromosomes with anaphasic precocity in the male mecopteran *Boreus brumalis* Fitch. J. Morph. **89**, 37—57.

COURT-BROWN, W. M., K. E. BUCKTON, P. A. JACOBS, I. M. TOUGH, E. V. KUENSSBERG, and J. D. E. KNOX, 1966: Chromosome studies on adults. Eugenics Lab. Memoirs **42**, 1—91.

CROUSE, H. V., 1965: Experimental alterations in the chromosome constitution of *Sciara*. Chromosoma (Berlin) **16**, 391—410.

DA CUNHA, A. B., and C. PAVAN, 1954: Duas novas configurações cromosômicas em *Tityus bahiensis* (Scorpiones-Buthidae). Ciencia e Cultura **6**, 18—20.

DAKER, M. G., 1966: "Kleine Liebling" a haploid cultivar of *Pelargonium*. Nature (London) **211**, 549—550.

D'AMATO, F., e M. G. AVANZI, 1948: Reazioni di natura auxinica ed effetti rizogeni in *Allium cepa* L. Studio cito-istologico sperimentale. Nuovo Giorn. Bot. Ital. ns**55**, 116—123.

Darlington, C. D., 1931: Chiasma formation and chromosome pairing in *Fritillaria*. Proc. 5th Int. Bot. Congr. Cambridge 189—191.
— 1939: The genetical and mechanical properties of the sex chromosomes V. *Cimex* and the *Heteroptera*. J. Genetics **39**, 101—137.
— 1956: Natural populations and the breakdown of classical genetics. Proc. Roy. Soc. (London) **B 145**, 350—364.
— 1963: Chromosome Botany. 2nd edition. Allen and Unwin, London.
— and L. F. Lacour, 1940: Differential reactivity of the chromosomes. Ann. Bot. N. S. **2**, 615—625.
— — 1941: The detection of inert genes. J. Hered. **52**, 115—121.
— and P. T. Thomas, 1941: Morbid mitosis and the activity of inert chromosomes in *Sorghum*. Proc. Roy. Soc. London **B 859**, 127—150.
— and E. K. Janaki-Ammal, 1945: Adaptive isochromosomes in *Nicandra*. Ann. Bot. N. S. **9**, 267—281.
— and P. C. Koller, 1947: The chemical breakage of chromosomes. Heredity (London) **1**, 187—222.
— and L. F. Lacour, 1950: Hybridity selection in *Campanula*. Heredity (London) **4**, 217—248.
— J. B. Hair, and R. Hurcombe, 1951: The history of the garden hyacinths. Heredity (London) **5**, 233—252.
— and A. P. Wylie, 1956: Chromosome atlas of flowering plants. Allen and Unwin, London.
— and G. W. Shaw, 1959: Parallel polymorphism in the heterochromatin of *Trillium* species. Heredity (London) **13**, 89—121.
Davies, E. W., 1956: Cytology, evolution and origin of the aneuploid series in the genus *Carex*. Hereditas **42**, 349—365.
Dearing, W. H. Jr., 1934: The material continuity and individuality of the somatic chromosomes of *Ambystoma tigrinum* with special reference to the nucleolus component. J. Morph. **56**, 157—179.
Dermen, H., 1941: Intranuclear polyploidy in bean induced by Naphthaleneacetic acid. J. Heredity **32**, 133—138.
— 1953: Periclinal cytochimeras and the origin of tissues in stem of peach. Amer. J. Bot. **40**, 154—168.
Dietz, R., 1958: Multiple Geschlechtschromosomen bei den cypriden Ostracoden, ihre Evolution und ihr Teilungsverhalten. Chromosoma (Berlin) **9**, 359—440.
Dolfini, S., and A. Gottardi, 1966: Changes of chromosome number in cells of *Drosophila melanogaster* cultured *in vitro*. Experientia (Basel) **22**, 144.
Duncan, R. E., and J. G. Ross, 1950: The nucleus in differentiation and development III. Nuclei of maize endosperm. J. Heredity **41**, 259—268.
Dyer, A. F., 1963: Allocyclic segments of chromosomes and the structural heterozygosity that they reveal. Chromosoma (Berlin) **13**, 545—576.
— 1964: Heterochromatin in American and Japanese species of *Trillium* I. Fusion of chromocentres and the distribution of H-segments Cytologia (Tokyo) **29**, 155—170. — II. The behaviour of H-segments. Cytologia (Tokyo) **29**, 171—190.

Edwards, J. H., D. G. Harnden, A. H. Cameron, V. M. Crosse, and O. H. Wolff, 1960: A new trisomic syndrome. Lancet **i**, 787—789.
Ehrendorfer, F., 1959: Unterschiedliche Störungssyndrome der Meiose bei diploiden und polyploiden Sippen des *Achillea-millefolium*-Komplexes und ihre Bedeutung für die Mikro-Evolution. Chromosoma (Berlin) **10**, 482—496.
— 1960: Akzessorische Chromosomen bei *Achillea:* Auswirkungen auf das Fortpflanzungssystem, Zahlen-Balance und Bedeutung für die Mikro-Evolution (Zur Phylogenie der Gattung *Achillea*, VI). Z. Vererbungslehre **91**, 400—422.
Einset, J., and B. Lamb, 1951: Chimeral sports of grapes. J. Heredity **42**, 159—162.
El-Alfi, O. S., P. M. Smith, and J. J. Biesele, 1965: Chromosome breaks in human leucocyte cultures induced by an agent in the plasma of infectious hepatitis patients. Hereditas (Lund) **52**, 285—294.
Ellis, J. R., R. Marshall, K. S. Normand, and L. S. Penrose, 1963: A girl with triploid cells. Nature (London) **198**, 411.
Elsdale, T. R., M. Fischberg, and S. Smith, 1958: A mutation that reduces nucleolar number in *Xenopus laevis*. Exp. Cell Res. **14**, 642—643.
Erbrich, P., 1965: Über Endopolyploidie und Kernstrukturen in Endospermhaustorien. Öst. Bot. Z. **112**, 197—262.

Estable, C., e J. R. Sotelo, 1951: Una nueva estructura celular: el nucleolonema. Inst. Inv. Cien. Biol. Publ. 1, 105—126.

— — 1955: The behaviour of the nucleolonema during mitosis. Pp. 170—190 in "Symposium on Fine Structure of cells." Noordhoff, Groningen.

Evans, G. M., A. Durrant, and H. Rees, 1966: Associated nuclear changes in the induction of flax genotrophs. Nature (London) 212, 697—699.

Evans, H. J., 1960: Supernumerary chromosomes in wild populations of the snail. *Helix pomatia* L. Heredity (London) 15, 129—138.

— 1961: The frequency and distribution of interchange and isochromatid aberrations induced by the irradiation of diploid and tetraploid cells. Pp. 259—270 in "Effects of ionising radiations on seeds." International Atomic Energy Agency, Vienna.

— and T. R. L. Bigger, 1961: Chromatid aberrations induced by gamma irridiation II. Non-randomness in the distribution of chromatid aberrations in relation to chromosome length in *Vicia faba* root tip cells. Genetics 46, 277—289.

— and V. Pond, 1964: The influence of the centromere on chromosome fragment frequency under chronic irradiation. Portugaliae Acta Biol. A 8, 125—146.

— C. E. Ford, M. F. Lyon, and J. Gray, 1965: DNA replication and genetic expression in female mice with morphologically distinguishable X chromosomes. Nature (London) 206, 900—903.

Fabergé, A. C., 1959: Production by alpha-particles of functionally stable broken chromosome ends in maize. Genetics 44, 279—285.

Fankhauser, G., 1945: The effects of changes in chromosome number on amphibian development. Quart. Rev. Biol. 20, 20—78.

Fedoroff, N., and R. Milkman, 1964: Specific puff induction by trytophan in *Drosophila* salivary gland chromosomes. Biol. Bull. 127, 369 (Abst.).

Ferguson-Smith, M. A., A. W. Johnston, and S. D. Handmaker, 1960: The chromosome complement in true hermaphroditism. Lancet ii, 126—128.

— — 1960: Primary amentia and micro-orchidism associated with an XXXY sex-chromosome constitution. Lancet ii, 184—177.

— W. S. Mack, P. M. Ellis, M. Dickson, B. Sanger, and R. R. Race, 1964: Parental age and the source of the X chromosomes in XXY Klinefelters syndrome. Lancet i, 46.

Fernandes, A., 1949: Le problème de l'hétérochromatinisation chez *Narcissus bulbocodium* L. Bol. Soc. Broteriana 23, 1—88.

Fischberg, M., and H. Wallace, 1960: A mutation which reduces nucleolar number in *Xenopus laevis.* Pp. 30—34 in "The Cell nucleus." Butterworth, London.

Fjelde, A., and O. A. Holtermann, 1962: Chromosome studies in HEp-2 tissue culture cell line during infection with measles virus. Life Sciences 12, 638—692.

Ford, C. E., 1960: Human cytogenetics: Its present place and future possibilities. Amer. J. Human Genet. 12, 104—117.

— 1964: Selection pressure in mammalian cell populations. Pp. 27—45 in "Cytogenetics of cells in culture." Symp. Int. Soc. Cell Biol. 3, Academic Press, London.

— J. L. Hamerton, and G. B. Sharman, 1957: Chromosome polymorphism in the common shrew. Nature (London) 180, 392—393.

— K. W. Jones, P. E. Polani, J. C. De Almeida, and J. H. Briggs, 1959: A sex chromosome anomaly in a case of gonadal dysgenesis (Turner's syndrome). Lancet i, 711—713.

— P. E. Polani, J. H. Briggs, and P. M. F. Bishop, 1959: A presumptive human XXY/XX mosaic. Nature (London) 183, 1030—1032.

— and E. P. Evans, 1964: A reciprocal translocation in the mouse between the X-chromosome and a short autosome. Cytogenetics 3, 295—305.

Ford, D. K., and G. Yerganian, 1958: Observations on the chromosomes of chinese hamster cells in tissue culture. J. Natl. Cancer Inst. 21, 393—425.

Fraccaro, M., A. I. Taylor, M. Bodian, and G. H. Newns, 1962: A human intersex ("true hermaphrodite") with XX/XXY/XXYYY sex chromosomes. Cytogenetics 1, 104—112.

Fredga, K., 1964: A new sex-determining mechanism in a mammal. Chromosomes of Indian mongoose (*Herpestes auropunctatus*). Hereditas (Lund) 52, 411—420.

Fröst, S., 1957: The inheritance of the accessory chromosomes in *Centaurea scabiosa.* Hereditas (Lund) 43, 403—422.

— 1960: A new mechanism for numerical increase of accessory chromosomes in *Crepis pannonica.* Hereditas (Lund) 46, 497—503.

— 1962: Numerical increase of accessory chromosomes in *Crepis conyzaefolia.* Hereditas (Lund) 48, 667—676.

GAGNON, J., N. KATYK-LONGTIN, J. A. DE GROOT, and A. BARBEAU, 1961: Double trisomie Autosomique à 48 chromosomes. Union Med. Canada **90**, 1220—1226.
GARTLER, S. M., S. H. WAXMAN, and E. GIBLETT, 1962: An XX/XY human hermaphrodite resulting from double fertilisation. Proc. Natl. Acad. Sci. U.S. **48**, 332—335.
GEITLER, L., 1939: Die Entstehung der polyploiden Somakerne der Heteropteren durch Chromosomenteilung ohne Kernteilung. Chromosoma (Berlin) **1**, 1—22.
— 1955: Riesenkerne im Endosperm von *Allium ursinum*. Öst. bot. Z. **102**, 460—475.
GERSTEL, D. E., and J. A. BURNS, 1965: Chromosomes of unusual length in hybrids between two species of *Nicotiana*. Pp. 41—56 in "Chromosomes Today, Vol. 1." Oliver and Boyd, Edinburgh.
GEYER-DUSZYNSKA, I., 1959: Experimental research on chromosome elimination in *Cecidomyidae* (*Diptera*). J. Exp. Zool. **141**, 391—448.
— 1961: Chromosome behaviour in spermatogenesis of *Cecidomyidae* (Diptera). Chromosoma (Berlin) **11**, 499—513.
GILBERT-DREYFUS, P., M. SEBAOUN-ZUGMAN, J. SEBAOUN, G. DELZANT et F. SCHAISON, 1963: Dégénérescence neurogerminale avec polysomie X et mosaïque complexe. Pathol. Biol. **2**, 1244—1246.
GILES, N., 1940: Spontaneous chromosome aberrations in *Tradescantia*. Genetics **25**, 69—87.
— 1941: Spontaneous chromosome aberrations in triploid *Tradescantia* hybrids. Genetics **26**, 632—649.
GIMÉNEZ-MARTIN, G., J. F. LÓPEZ-SÁEZ, and A. GONZÁLEZ-FERNÁNDEZ, 1963: Somatic chromosome structure (Observations with the light microscope). Cytologia (Tokyo) **28**, 381—389.
GOLDSCHMIDT, E., 1953: Multiple sex-chromosome mechanisms and polyploidy in animals. J. Genet. **51**, 434—440.
GOODRICH, H. B., 1914: The maturation divisions in *Ascaris incurva*. Biol. Bull. **27**, 147—150.
— 1916: The germ cells in *Ascaris incurva*. J. Exp. Zool. **21**, 61—100.
GORMAN, G. C., and L. ATKINS, 1966: Chromosomal heteromorphism in some male lizards of the genus *Anolis*. Amer. Nat. **100**, 579—583.
— — and T. HOLZINGER, 1967: New karyotypic data on 15 genera of lizards in the family *Iguanidae* with a discussion of traxonomic and cytologic implications. Cytogenetics **6**, 286—299.
GURDON, J. B. 1959: Tetraploid frogs. J. Exp. Zool. **141**, 519—543.
GRANT, V., 1963: The origin of adaptations. Columbia Univ. Press, New York.
GREENLEAF, W. H., 1938: Induction of polyploidy in *Nicotiana*. J. Heredity **29**, 451—464.
GRIFFEN, A. B., and M. C. BUNKER, 1964: Three cases of trisomy in the mouse. Proc. Natl. Acad. Sci. U.S. **52**, 1194—1198.
GUÉNIN, H. A., 1948: La formule chromosomique de *Blaps mortisaga* L. (Coleoptera, Tenebrionidae). Experientia **4**, 221.
— 1950: Chromosomes et hétérochromosomes de Tenebrionides. Genetica **25**, 157—182.
— 1952: Hétérochromosomes de Cicindèles. Rev. Suisse Zool. **59**, 277—282.
GUSTAFSSON, A., 1947: Apomixis in higher plants III. Biotype and species formation. Acta Univ. Lund, N.F. Avd. **44**, 183—370.
GUSTAVSSON, I., 1966: Chromosome abnormality in cattle. Nature (London) **211**, 865—866.
— and G. ROCKBORN, 1964: Chromosome abnormality in three cases of lymphatic leukaemia in cattle. Nature (London) **203**, 990.

HAGA, T., 1961: Intra-individual variation in number and linear patterning of the chromosomes I. B-chromosomes in *Rumex, Paris* and *Scilla*. Proc. Japan. Acad. **37**, 627—632.
— and M. KURABAYASHI, 1954: Genom and polyploidy in the genus *Trillium* V. Chromosomal variation in natural population of *Trillium kamtschaticum* Pall. Mem. Fac. Sci. Kyushu Univ. **El**, 159—185.
HAIR, J. B., and E. J. BEUZENBERG, 1958: Chromosomal evolution in the Podocarpaceae. Nature (London) **181**, 1584—1586.
HÅKANSSON, A., 1956: Seed development of *Brassica oleracea* and *B. rapa* after certain reciprocal pollinations. Hereditas (Lund) **42**, 373—396.
— 1957: Notes on the giant chromosomes of *Allium nutans*. Bot. Notis. **110**, 196—200.
— and A. LEVAN, 1942: Nucleolar conditions in *Pisum*. Hereditas (Lund) **28**, 436—440.

HALKKA, O., 1964: A photometric study of the *Luzula* problem. Hereditas (Lund) **52**, 81—88.

HAMPAR, B., and S. A. ELLISON, 1961: Chromosomal aberrations induced by an animal virus. Nature (London) **192**, 145—147.

— — 1963: Cellular alterations in the MCH line of chinese hamster cells following infection with Herpes simplex virus. Proc. Natl. Acad. Sci. U.S. **49**, 474—480.

HANNAH, A., 1951: Localisation and function of heterochromatin in *Drosophila melanogaster*. Adv. Genet. **4**, 87—125.

HARNDEN, D. G., 1964: Cytogenetic studies on patients with virus infections and subjects vaccinated against yellow fever. Amer. J. Human Genetics **16**, 201—213.

HASITSCHKA, G., 1956: Bildung von Chromosomenbündeln nach Art der Speicheldrüsenchromosomen. Spiralisierte Ruhekernchromosomen und andere Struktureigentümlichkeiten in den endopolyploiden Riesenkernen der Antipoden von *Papaver rhoeas*. Chromosoma (Berlin) **8**, 87—113.

HASITSCHKA-JENSCHKE, G., 1958: Zur Karyologie der Samenanlage dreier *Allium*-Arten. Öst. bot. Z. **105**, 71—72.

— 1959: Vergleichende karyologische Untersuchungen an Antipoden. Chromosoma (Berlin) **10**, 229—267.

HAUSCHKA, T. S., and V. V. BRUNST, 1964: Sexual dimorphism in the nucleolar autosome of the axolotl (*Siredon mexicanum*). Hereditas (Lund) **52**, 345—356.

HAUSCHTECK, E., 1962: Die Zytologie der Pädogenese und der Geschlechtsbestimmung einer heterogenen Gallmücke. Chromosoma (Berlin) **13**, 163—182.

HAYMAN, D. L., and P. G. MARTIN, 1965: Sex chromosome mosaicism in the marsupial genera *Isoodon* and *Perameles*. Genetics **52**, 1201—1206.

HELWIG, E. R., 1941: Multiple chromosomes in *Philocleon anomalus* (Orthoptera: Acrididae). J. Morph. **69**, 317—327.

HENEEN, W., K., 1963: Extensive chromosome breakage occurring spontaneously in a certain individual of *Elymus farctus* (= *Agropyron junceum*). Hereditas (Lund) **49**, 1—32.

HENKING, H., 1891: Untersuchungen über die ersten Entwicklungsvorgänge in den Eiern der Insekten II. Z. wiss. Zool. **51**, 685—736.

HENNEN, S., 1963: Chromosomal and embryological analyses of nuclear changes occurring in embryos derived from transfers of nuclei between *Rana pipiens* and *Rana sylvatica*. Develop. Biol. **6**, 133—183.

HENRICSON, B., and L. BACKSTRÖM, 1964: Translocation heterozygosity in a boar. Hereditas (Lund) **52**, 166—170.

HEWITT, G. M., and B. JOHN, 1965: The influence of numerical and structural chromosome mutations on chiasma conditions. Heredity (London) **20**, 123—135.

— and G. SCHROETER, 1968: Compound polymorphism in *Oedaleonotus enigma* I. The karyotypic facies (in preparation).

HIRSCHHORN, K., W. H. DECKER, and H. L. COOPER, 1960: Human intersex with chromosome mosaicism of the type XY/XO: Report of a case. New Engl. J. Med. **263**, 1044—1048.

HSU, T. C., 1954: Cytological studies on HeLa, a strain of human cervical carcinoma I. Observations on mitosis and chromosomes. Texas Rep. Biol. Med. **12**, 833—846.

HUGHES-SCHRADER, S., 1948: Cytology of Coccids (*Coccoidea, Homoptera*). Adv. Genet. **2**, 127—203.

— 1950: The chromosomes of mantids (*Orthoptera, Manteidae*) in relation to taxonomy. Chromosoma (Berlin) **4**, 1—55.

— 1957: Differential polyteny and polyploidy in diaspine coccids (*Homoptera: Coccoidea*). Chromosoma (Berlin) **8**, 709—718.

— 1958: The DNA content of the nucleus as a tool in the cytotaxonomic study of insects. Proc. Xth Int. Congr. Entomol. **2**, 935—944.

— and H. RIS, 1941: The diffuse spindle of coccids, verified by the mitotic behaviour of induced chromosome fragments. J. Exp. Zool. **87**, 426—456.

— and F. SCHRADER, 1956: Polyteny as a factor in the chromosomal evolution of the Pentatomini (*Hemiptera*). Chromosoma (Berlin) **8**, 135—151.

HUMPHREY, R. R., and G. FRANKAUSER, 1946: The ovaries in triploid axolotl females of different genotypes with respect to sex chromosomes. Anat. Rec. **94**, 472 (Abstract).

JACOBS, P. A., and J. A. STRONG, 1959: A case of human intersexuality having a possible XXY sex-determining mechanism. Nature (London) **183**, 302—303.

JACOBS, P. A., D. G. HARNDEN, W. M. COURT-BROWN, J. GOLDSTEIN, H. G. CLOSE, T. N. MACGREGOR, N. MACLEAN, and J. A. STRONG, 1960: Abnormalities involving the X-chromosome in women. Lancet i, 1213—1216.
— — K. E. BUCKTON, W. M. COURT-BROWN, M. J. KING, J. A. MCBRIDE, T. N. MACGREGOR, and N. MACLEAN, 1961: Cytogenetics studies in primary amenorrhoea. Lancet i, 1183—1189.
JACOBSEN, P., 1957: The sex chromosomes in *Humulus* L. Hereditas (Lund) **43**, 357—370.
JACKSON, R. C., 1962: Interspecific hybridisation in *Haplopappus* and its bearing on chromosome evolution in the Blepharodon section. Amer. J. Bot. **49**, 119—132.
— 1965: A cytogenetic study of a three paired race of *Haplopappus gracilis*. Amer. J. Bot. **52**, 946—953.
JAIN, H. K., 1957: Effect of high temperature on meiosis in *Lolium:* nucleolar inactivation. Heredity (London) **11**, 23—26.
JANAKI-AMMAL, E. K. 1958: Iso-chromosomes and the origin of triploidy in hybrids between old and new world species of *Philadelphus*. Proc. Indian Acad. Sci. **43**, 251—258.
JOHN, B., 1957: The chromosomes of zooparasites I. *Acanthocephalus ranae. (Acanthocephala: Echinorhynchidae)*. Chromosoma (Berl.) **8**, 730—738.
— II. *Oswaldocruzia filiformis* (*Nematoda: Trichostrongylidae*). Chromosoma (Berlin) **9**, 61—68.
— and K. R. LEWIS, 1959: Selection for interchange heterozygosity in an inbred culture of *Blaberus discoidalis*. Genetics **44**, 251—267.
— and S. A. HENDERSON, 1962: Asynapsis and polyploidy in *Schistocerca paranensis*. Chromosoma (Berlin)**13**, 111—147.
— and G. M. HEWITT, 1963: A spontaneous interchange in *Chorthippus brunneus* with extensive chiasma formation in an interstitial segment. Chromosoma (Berlin) **14**, 638—650.
— — 1965: The B-chromosome system of *Myrmeleotettix maculatus* (Thunb.) I. The mechanics. Chromosoma (Berlin) **16**, 548—578. — II. The statics. Chromosoma (Berlin) **17**, 121—138.
— and K. R. LEWIS, 1965: The meiotic system. Protoplasmatologia **VI, F1**, 1—335, Springer-Verlag Wien.
— and G. M. HEWITT, 1966 a: A polymorphism for heterochromatic supernumerary segments in *Chorthippus parallelus*. Chromosoma (Berlin) **18**, 254—271.
— — 1966 b: Karyotype stability and DNA variability in the acrididae. Chromosoma (Berlin) **20**, 155—172.
— and K. R. LEWIS, 1966: Chromosome variability and geographic distribution in insects. Science **152**, 711—721.
— and D. D. SHAW, 1967: Karyotype variation in dermestid beetles. Chromosoma (Berlin) **20**, 371—385.
JONES, K., 1957: Some aspects of plant variation in the grasses. Pp. 45—55 in "Progress in the study of the British flora." Buncle & Co., Arbroath.
— 1964: Chromosomes and the nature and origin of *Anthoxanthum odoratum* L. Chromosoma (Berlin) **15**, 248—274.

KAHN, J., 1962: The nucleolar organiser in the mitotic chromosome complement of *Xenopus laevis*. Quart. J. Micr. Sci. **103**, 407—409.
KAWAMURA, T., 1951: Reproductive ability of triploid newts with remarks on their offspring. J. Sci. Hiroshima Univ. Bl. **12**, 1—10.
KAYANO, H., 1960: Chiasma studies in structural hybrids III. Reductional and equational separation in *Disporum sessile*. Cytologia (Tokyo) **25**, 461—467.
KEUNEKE, W., 1924: Über die Spermatogenese einiger Dipteren. Z. Zellenlehre **1**, 357—412.
KEYL, H. G., 1960: Die cytologische Diagnostik der Chironomiden II. Diagnosen der Geschwisterarten *Chironomus acidophilus* n.sp. and *Ch. uliginosus* n.sp. Arch. Hydrobiol. **57**, 187—195.
— 1962: Chromosomenevolution bei *Chironomus* II. Chromosomenbauten und phylogenetische Beziehungen der Arten. Chromosoma (Berlin) **13**, 464—514.
— 1965: A demonstrable local and geometric increase in the chromosomal. DNA of Chironomus. Experientia **21**, 191.
— 1966: Increase of DNA in chromosomes. Pp. 99—101 in "Chromosomes Today." Vol. 1. Oliver and Boyd, Edinburgh.
KICHIJO, H., 1942: Chromosomes of *Oligotoma japonica* (Embioptera). Jap. J. Genet. **18**, 196—197.

KIMURA, M., 1962: A suggestion on the experimental approach to the origin of supernumerary chromosomes. Amer. Nat. **96**, 319—320.
— and H. KAYANO, 1961: The maintenance of supernumerary chromosomes in wild populations of *Lilium callosum* by preferential segregation. Genetics **46**, 1699—1712.
KLINGSTEDT, H., 1939: Taxonomic and cytological studies on grasshopper hybrids. J. Genetics **37**, 389—420.
KOBEL, H. R., 1962: Heterochromosomen bei *Vipera berus* L. (Viperidae, Serpentes). Experientia **18**, 173—174.
KOO, F. K. S., 1958: Pseudo-isochromosomes produced in *Avena strigosa* Schreb by ionising radiations Cytologia (Tokyo) **23**, 109—111.
KRACZKIEWICZ, Z., 1937: Recherches cytologiques sur le cycle de *Miastor metraloas*. Cellule **46**, 57—74.
KRISHAN, A., G. J. HAIDEN, and R. N. SHOFFNER, 1965: Mitotic chromosomes and the W-sex chromosome of the great horned owl (*Bubo v. virginianus*). Chromosoma (Berlin) **17**, 258—263.
— and R. N. SHOFFNER, 1966: Sex chromosomes in the domestic fowl. (*Gallus domesticus*), turkey (*Meleagris gallopavo*) and the chinese pheasant (*Phasianus colchicus*). Cytogenetics **5**, 53—63.
KROEGER, H., 1960: The induction of new puffing patterns by transplantation of salivary gland nuclei into egg cytoplasm of *Drosophila*. Chromosoma (Berlin) **11**, 129—145.
— 1963: Experiments on the extra-nuclear control of gene activity in dipteran polytene chromosomes. J. Cell. Comp. Physiol. **62**, Suppl. 1, 45—59.
— 1966: Potentialdifferenz und Puffmuster. Elektrophysiologische und cytologische Untersuchungen an den Speicheldrüsen von *Chironomus thummi*. Exp. Cell Res. **41**, 64—80.
KURITA, M., 1964: Chromosome study in *Nothoscordum inutile*. Bot. Mag. (Tokyo) **77**, 81—85.
KYHOS, D. W., 1965: The independent aneuploid origin of two species of *Chaenactis* (Compositae) from a common ancestor. Evolution **19**, 26—43.

LA COUR, L. F., 1966: The internal structure of nucleoli. Pp. 150—160 in "Chromosomes today." Vol. 1. Oliver and Boyd, Edinburgh.
LE CALVEZ, J., 1949: Données caryologiques sur l'Embioptère *Monotylota ramburi* Endere. C. R. Acad. Sci. (Paris) **229**, 245—246.
LEJEUNE, J., M. GAUTIER et R. TURPIN, 1959: Étude des chromosomes somatique de neuf enfants Mongolien. C. R. Acad. Sci. (Paris) **248**, 1721—1725.
LEVAN, A., 1933: Cytological studies in *Allium fistulosum*. Svensk. Bot. Tidskrift. **27**, 211—232.
— 1966: Non-random representation of chromosome types in human tumor stemlines. Hereditas (Lund) **55**, 28—38.
— and S. L. EMSWELLER, 1938: Structural hybridity in *Northoscordum fragrans*. J. Heredity **29**, 291—294.
— and T. S. HAUSCHKA, 1953: Endomitotic reduplication mechanisms in ascites tumours of the mouse. J. Natl. Cancer Inst. **14**, 1—43.
— and J. J. BIESELE, 1958: Role of chromosomes in cancerogenesis as studied in serial tissue culture of mammalian cells. Ann. N.Y. Acad. Sci. **71**, 1022—1053.
— T. C. HSU, and H. F. STICH, 1962: The idiogram of the mouse. Hereditas (Lund) **48**, 673—687.
— K. FREDGA, and A. A. SANDBERG, 1964: Nomenclature for centromeric position on chromosomes. Hereditas (Lund) **52**, 201—220.
LEVITAN, M., 1966: Evidence for chromosomal properties of a nucleolus. Genetics **54**, 345—346.
LEWIS, H., 1951: The origin of supernumerary chromosomes in natural populations of Clarkia elegans. Evolution **5**, 142—157.
— 1962: Catastrophic selection as a factor in speciation. Evolution **16**, 257—271.
LEWIS, K. R., 1958: Chromosome structure and organisation in *Pellia epiphylla*. Phyton **11**, 29—37.
— and B. JOHN, 1957: The organisation and evolution of the sex multiple in *Blaps mucronata*. Chromosoma (Berlin) **9**, 69—80.
— — 1959: Breakdown and restoration of chromosome stability following inbreeding in a locust. Chromosoma (Berlin) **10**, 589—618.
— — 1963: Spontaneous interchange in *Chorthippus brunneus*. Chromosoma (Berlin) **14**, 618—637.

Lewis, K. R., and B. John, 1963: "Chromosome Marker." Churchill, London.
Lewis, W. H., 1962: Aneusomaty in aneuploid populations of *Claytonia virginica.* Amer. J. Bot. **19**, 918—928.
Lezzi, M., 1966: Induktion eines Ecdyson-aktivierbaren Puff in isolierten Zellkernen von *Chironomus* durch KCl. Exp. Cell Res. **43**, 571—577.
Li, N., and R. C. Jackson, 1961: Cytology of supernumerary chromosomes in *Haplopappus spinulosus* ssp. cotula. Amer. J. Bot. **48**, 419—426.
Lima-de-Faria, A., 1956: The role of the kinetochore in chromosome organisation. Hereditas (Lund) **42**, 85—160.
— 1963: The evolution of the structural pattern in a rye B-chromosome. Evolution **17**, 289—295.
— and P. Sarvella, 1958: The organisation of telomeres in species of *Solanum, Salvia, Scilla, Secale, Agapanthus* and *Ornithogalum.* Hereditas **44**, 337—346.
— — 1962: Variation of the chromosome phenotype in *Zea, Solanum* and *Salvia.* Chromosoma (Berlin) **13**, 300—314.
— — and R. Morris, 1959: Different chromomere numbers at meiosis and mitosis in *Ornithogalum.* Hereditas (Lund) **45**, 467—480.
Linnert, G., 1965: Untersuchungen an polyhaploiden Nachkommen Autotetraploider II. Die spontane Entstehung von Genom-, Chromosomen- und Genmutationen während des tetraploiden Zustandes bei *Oenothera hookeri* de V. Z. Vererbungslehre **96**, 190—212.
Löve, A., 1943: Cytogenetic studies on *Rumex* subgenus *Acetosella.* Hereditas (Lund) **30**, 1—136.
— and N. Sarkar, 1956: Cytotaxonomy and sex determination in *Rumex paucifolius.* Canad. J. Bot. **34**, 261—268.
Löve, D., 1944: Cytogenetic studies on dioecious *Melandrium.* Bot. Notis. **97**, 125—214.
Lutman, B. S., 1934: Cell size and structure in plants as affected by various inorganic elements. Vt. Agr. Exp. Sta. Bull. **383**, 1—54.
Lyon, M. F., and R. Meredith, 1966: Autosomal translocations causing male sterility and viable aneuploidy in the mouse. Cytogenetics **5**, 335—354.

McClintock, B., 1931: Cytological observations on deficiencies involving known genes, translocations and an inversion in *Zea mays.* Miss. Agr. Exp. Sta. Res. Bull. **163**, 1—30.
— 1941: The stability of broken ends of chromosomes in *Zea mays.* Genetics **26**, 234—282.
McClung, C. E., 1902: The spermatocyte divisions of the locustidae. Kans. Univ. Sci. Bull. **1**, 185—238.
McFee, A. F., M. W. Banner, and J. M. Rary, 1966: Variation in chromosome number among European wild pigs. Cytogenetics **5**, 75—81.
MacLean, N., D. G. Harnden, W. M. Court-Brown, J. Bond, and D. J. Mantle, 1964: Sex chromosome abnormalities in newborn babies. Lancet **i**, 286—290.
— J. M. Mitchell, D. G. Harnden, J. Williams, P. A. Jacobs, K. A. Buckton, A. G. Baikie, W. M. Court-Brown, J. A. McBridge, J. A. Strong, H. G. Close, and D. C. Jones, 1962: A survey of sex chromosome abnormalities among 4514 mental defectives. Lancet **i**, 293—296.
Makino, S., 1951: An atlas of the chromosome number in animals. Iowa State College Press. Iowa.
— K. Yamada, and T. Kajii, 1965: Chromosome aberrations in leucocytes of patients with aseptic meningitis. Chromosoma **16**, 372—380.
Manna, G. K., 1958: Cytology and inter-relationships between various groups of Heteroptera. Proc. Xth. Int. Entomol. congress **2**, 919—934.
— 1962: A further evaluation of the cytology and inter-relationships between various groups of Heteroptera. The Nucleus **5**, 7—28.
— and S. G. Smith, 1959: Chromosome polymorphism and inter-relationships among bark weevils of the genus *Pissodes* Germar. The Nucleus **2**, 179—208.
Marsden, P. M., D. W. Smith, and M. J. McDonald, 1964: Congenital abnormalities in the newborn infant including minor variations. J. Paediatrics **64**, 357—371.
Markaryan, D., and J. Schultz-Schaffer, 1958: A possible origin of supernumerary fragment chromosomes. J. Heredity **9**, 3—7.
Marks, G. E., 1957: The cytology of *oxalis dispar.* Chromosoma (Berlin) **8**, 650—670.
Martin, F. W., 1966: Sex ratio and sex determination in *Dioscorea.* J. Heredity **57**, 95—99.

MARTIN, P. G., and D. L. HAYMAN, 1966: Complex sex-chromosome system in the hare-wallaby *Lagorchestes conspicillatus* Gould. Chromosoma (Berlin) **19**, 159—175.

— and R. SHRANKS, 1966: Does *Vicia faba* have multi-stranded chromosomes? Nature (London) **211**, 650—651.

MATHER, K., 1932: Chromosome variation in *Crocus* I. J. Genet. **26**, 129—142.

MATTHEY, R., 1931: Chromosomes de reptiles sauriens, ophidiens et cheloniens. L'évolution de la formule chromosomiale chez les sauriens. Rev. suisse Zool. **38**, 117—186.

— 1933: Nouvelle contribution à l'étude des chromosomes chez les sauriens. Rev. suisse Zool. **40**, 281—318.

— 1963 a: Polymorphisme chromosomique intraspecifique et intra individuel chez *Acomys minous* ♂ × *Acomys cahirinus* ♀. Le mécanisme des fusions centriques. Chromosoma (Berlin) **14**, 468—497.

— 1963 b: Cytologie comparée et polymorphisme chromosomique chez des *Mus* africains appartenant aux groupes *bufo-triton* et *minutoides* (Mammalia-Rodentia). Cytogenetics **2**, 290—322.

— 1964: Evolution chromosomique et spéciation chez les *Mus* du sous-genre *Leggada* Gray 1837. Experientia *20*, 657—712.

— 1965 a: Le problème de la détermination du sexe chez *Acomys Selousi* de Winton Cytogénétique du genre *Acomys* (Rodentia murinae). Rev. suisse Zool. **72**, 119—144.

— 1965 b: Un type nouveau de chromosomes sexuels multiples chez une souris africaine du groupe *Mus* (*Leggada*) *minutoides* (Mammalia Rodentia). Male: X_1X_2/Y. Femelle: X_1X_2/X_1X_2. Chromosoma (Berlin) **16**, 351—364.

— 1966: Une inversion péricentrique à l'origine d'un polymorphisme chromosomique non-Robertsonien dans une population de *Mastomys* (Rodentia-Murinae). Chromosoma (Berlin) **18**, 188—200.

— and J. AUBERT, 1947: Les chromosomes des Plecoptères. Bull. Biol. **81**, 202—246.

MATTSON, Ø., 1963: Telocentric chromosomes in *Tradescantia commelinoides*. Bot. Tidskrift **59**, 195—208.

MATUSZEWSKI, B., 1965: Transition from polyteny to polyploidy in salivary glands of Cecidomyiidae. Chromosoma (Berlin) **16**, 22—34.

MAUDE, P. F., 1940: Chromosome numbers in some British plants. New Phytol. **39**, 17—32.

MAZZONE, H. M., and G. YERGANIAN, 1963: Gross and chromosomal cytology of virus infected Chinese hamster cells. Exp. Cell Res. **30**, 591—592.

MECHELKE, F., 1955: Temperaturbedingte Chromosomensegmentierung bei Sommer- und Wintergersten. Kulturpfl. **3**, 127—136.

MELANDER, Y., 1950: Accessory chromosomes in animals especially in *Polycelis tenuis*. Hereditas (Lund) **36**, 261—296.

MELLO-SAMPAYO, T., 1961: Differential polyteny and karyotype evolution in *Luzula*, a critical interpretation of morphological and cytophotometric data. Genetica Iberica **13**, 1—22.

MESA, A., 1960: Cariologia de una nueva especie Uruguya del genero *Scottusa* (Orthoptera, Catantopidae). Rev. Soc. Urug. Ent. **4**, 87—94.

MEYLAN, A., 1965: Repartition géographiques des races chromosomiques de "*Sorex araneus*" L. en Europe (Mammalia-Insectivora). Rev. suisse Zool. **72**, 636—646.

MIKAMO, K., and E. WITSCHII, 1966: The mitotic chromosomes in *Xenopus laevis* (Daudin): Normal, sex reversed and female WW. Cytogenetics **5**, 1—19.

MOENS, P. B., 1965: The transmission of a heterochromatic isochromosome in *Lycopersicon esculentum*. Canad. J. Genetics and Cytology **7**, 296—303.

MOORHEAD, P. S., and E. SAKSELA, 1965: The sequence of chromosome aberrations during SV_{40} transformation of a human diploid cell strain. Hereditas (Lund) **52**, 271—284.

MORESCALCI, A., 1963: Conferma della presenza di eterocromosomi in *Xenopus laevis* Daudin. Rend. Acc. Sci. (Napoli) **Ser. 4, 30**, 310—314.

— 1964: Il corredo cromosomico di *Discoglossus pictus* Otth.: cromosomi sessuali, spiralizzazione cromosomica e zone eterocromatiche. Caryologia **17**, 327—345.

MORGAN, W. P., 1928: A comparative study of the spermatogenesis of five species of earwigs. J. Morph. **46**, 241—273.

MORRIS, R., 1955: Induced reciprocal translocations involving homologous chromosomes in maize. Amer. J. Bot. **42**, 546—550.

MORRISON, J. W. 1954: Chromosome interchange by misdivision in *Triticum*. Canad. J. Bot. **32**, 281—284.

Mukerjee, B. B., and A. K. Sinha, 1964: Single-active-X hypothesis: Cytological evidence for random inactivation of X-chromosomes in a female mule complement. Proc. Natl. Acad. Sci. **51**, 252—259.
Muldal, S., 1952: The chromosomes of the earthworms. I. The evolution of polyploidy. Heredity (London) **6**, 55—76.
— and C. H. Ockey, 1960: The "double male" a new chromosome constitution in Kleinefelter's syndrome. Lancet **ii**, 492—493.
Muller, H. J., 1932: Further studies in the nature and causes of gene mutation. Proc. VIth Int. Congr. Genet. **1**, 213—255.
— 1940: Bearings of the "*Drosophila*" work on systematics. Pp. 185—268 in "The New Systematics." Claredon Press; Oxford.
— and I. H. Herskowitz, 1954: Concerning the healing of chromosome ends produced by breakage in *Drosophila melanogaster*. Amer. Nat. **LXXXVIII**, 177—208.
Mülsow, 1911: Chromosomenverhältnisse bei *Ancryracanthus*. Zool. Anz. **38**, 484—486.
Müntzing, A., 1948: Acessory chromosomes in *Poa alpina*. Heredity (London) **2**, 49—61.
— 1954: Cyto-genetics of accessory chromosomes (B-chromosomes). Caryologia **6**, suppl., 282—301.

Nagl, W., 1965: Die Sat-Riesenchromosomen der Kerne des Suspensors von *Phaseolus coccineus* und ihr Verhalten während der Endomitose. Chromosoma (Berlin) **16**, 511—520.
Navashin, M., 1926: Variabilität des Zellkerns bei *Crepis*-Arten in bezug auf die Artbildung. Z. Zellforsch u. Mikr. Anat. **4**, 171—215.
— 1934: Chromosome alterations caused by hybridisation and their bearing upon certain general genetic problems. Cytologia (Tokyo) **5**, 169—203.
Newcomer, E. H., 1957: The mitotic chromosomes of the domestic fowl. J. Heredity **XLVIII**, 227—234.
— 1959: The meiotic chromosomes of the fowl. Cytologia (Tokyo) **24**, 403—410.
Nichols, W. W., 1963: Relationships of viruses, chromosomes and carcinogenesis. Hereditas (Lund) **50**, 53—80.
— 1966: Studies on the role of viruses in somatic mutation. Hereditas (Lund) **55**, 1—27.
— A. Levan, B. Hall, and G. Östergren, 1962: Measles associated chromosome breakage. Preliminary communication. Hereditas (Lund) **48**, 367—370.
— — P. Aula, and E. Norby, 1964: Extreme chromosome breakage induced by measles virus in different *in vitro* systems. Preliminary communication. Hereditas (Lund) **51**, 380—382.
— — L. L. Coriell, H. Goldner, and C. G. Ahlstrom, 1964: *In vitro* chromosome abnormalities in human leukocytes associated with Schmidt-Ruppin Rous sarcoma virus. Science **146**, 248—250.
— — P. Aula, and E. Norby, 1965: Chromosome damage associated with the measles virus *in vitro*. Hereditas (Lund) **54**, 101—118.
Nicklas, R. B., 1959: An experimental and descriptive study of chromosome elimination in *Miastor* spec. (Cecidomyidae; Diptera). Chromosoma (Berlin) **10**, 301—336.
— 1960: The chromosome cycle of a primitive cecidomyiid — *Mycophila speyeri*. Chromosoma (Berlin) **11**, 402—418.
Niyama, H., 1956: Cytological demonstration of an XO sex-mechanism in males of *Tecticeps japonicus*, an isopod Crustacea. Cytologia (Tokyo) **21**, 38—43.
Noda, S., 1960: Chiasma studies in structural hybrids II. Reciprocal translocation in *Lilium Maximowiczii*. Cytologia (Tokyo) **25**, 456—460.
Nordenskiold, H., 1951: Cyto-taxonomical studies in the genus *Luzula* I. Somatic chromosomes and chromosome numbers. Hereditas (Lund) **37**, 323—355.
— 1961: Tetrad analyses and the course of meiosis in three hybrids of *Luzula campestris*. Hereditas (Lund) **47**, 203—238.
Nowell, P. C., and D. A. Hungerford, 1960: Chromosome studies on normal and leukaemic human leucocytes. J. Natl. Cancer Inst. **25**, 85—93.
Núñez, O., 1962: Cytology of Collembola. Nature (London) **194**, 946—947.
Nur, U., 1961: Meiotic behaviour of an unequal bivalent in the grasshopper *Calliptamus palaestinensis* Bdhr. Chromosoma (Berlin) **12**, 272—279.

NUR, U., 1963: A mitotically unstable supernumerary chromosome with an accumulation mechanism in a grasshopper. Chromosoma (Berlin) **14**, 407—422.

— 1965: A modified Comstockiella chromosome system in the olive scale insect *Parlatoria oleae* (Coccoidea: Diaspididae). Chromosoma (Berlin) **17**, 104—120.

NYGREN, A., 1957: *Poa timoleontis* Heldr., a new diploid species in the section Bolbophorum A. and Gr. with accessory chromosomes only in the meiosis. Kungl. Lantbrukshogskolans Ann. **23**, 489—495.

OGAWA, K., 1952: Chromosome studies in the *Myriapoda VI*. A study on the sex chromosomes in two allied species of chilopods. Annot. Zool. Jap. **25**, 434—440.

— 1962: Unusual features of chromosomes found in *Thereuonema hilgendorfi* (Chilopoda). C. I. S. (Tokyo) **3**, 5.

OHNO, S., and R. KINOSITA, 1955: The primary and secondary constrictions on the chromosomes of the rat lymphoblast. Exp. Cell Res. **8**, 558—562.

— W. D. KAPLAN, and R. KINOSITA, 1959: Do XY- and O-sperm occur in *Mus musculus*? Exp. Cell Res. **18**, 382—384.

— E. T. KOVACS, and R. KINOSITA, 1960: A Robertsonian type of chromosomal change in L 4946 mouse ascites lymphoma. J. Natl. Cancer Inst. **24**, 1187—1197.

— C. WEILER, and C. STENIUS, 1961: A dormant nucleolus organizer in the guinea pig, *Cavia cobaya*. Exp. Cell Res. **25**, 498—503.

— J. M. TRUJILLO, W. D. KAPLAN, R. KINOSITA, and C. STENIUS, 1961: Nucleolus organizers in the causation of chromosomal anomalies in man. The Lancet **i**, 123—126.

— and B. M. CATTANACH, 1962: Cytological study of an X-autosome translocation in *Mus musculus*. Cytogenetics **1**, 129—140.

— L. C. CHRISTIAN, and C. STENIUS, 1962: Nucleolus organising microchromosomes of *Gallus domesticus*. Exp. Cell Res. **27**, 612—614.

— W. A. KITTRELL, L. C. CHRISTIAN, C. STENIUS, and G. A. WITT, 1963: An adult triploid chicken (*Gallus domesticus*) with a left ovotestis. Cytogenetics **2**, 42—49.

— J. JAINCHILL, and C. STENIUS, 1963: The creeping vole (*Microtus oregoni*) as a gonosomic mosaic I. The OY/XY constitution of the male. Cytogenetics **2**, 232—239.

— W. BECAK, and M. L. BECAK, 1964: X-autosome ratio and the behaviour pattern of individual X-chromosomes in placental mammals. Chromosoma (Berlin) **15**, 14—30.

— and M. F. LYON, 1965: Cytological study of Searle's X-autosome translocation in *Mus musculus*. Chromosoma (Berlin) **16**, 90—100.

— C. STENIUS, E. FAISST, and M. T. ZENZES, 1965: Post zygotic chromosomal rearrangements in rainbow trout (*Salmo irideus* Gibbons). Cytogenetics **4**, 117—119.

— and N. B. ATKINS, 1966: Comparative DNA values and chromosome complements of eight species of fishes. Chromosoma (Berlin) **18**, 455—466.

— C. WEILER, J. POOLE, L. CHRISTIAN, and C. STENIUS, 1966: Autosomal polymorphism due to pericentric inversions in the deer mouse (*Peromyscus maniculatus*) and some evidence of somatic segregation. Chromosoma (Berlin) **18**, 177—187.

— C. STENIUS, and L. CHRISTIAN, 1966: The XO as the normal female of the creeping vole (*Microtus oregoni*). Pp. 182—187 in "Chromosomes Today." Vol. 1. Oliver and Boyd, Edinburgh.

— 1968: Gene duplication by ancient polyploidy in vertebrate evolution (in press, "Chromosomes Today," Vol. 2).

OKSALA, T. Zytologische Studien an Odonaten I (1943) — Chromosomenverhältnisse bei der Gattung *Aeschna* mit besonderer Berücksichtigung der postreduktionellen Teilung der Bivalente. Ann. Acad. Sci. Fenn. **A 4**, 1—65: II (1944) — Die Entstehung der meiotischen Präkozität. Ann. Acad. Sci. Fenn. **A 4**, 1—37: III (1945) Die Ovogenese. Ann. Acad. Sci. Fenn. **A 4**, 1—32.

OLIVER, J. H.: Cytogenetics of Ticks (Acari: Ixodoidea) I (1966). Karyotypes of two *Ornithodoros* spezies (Argasidae) restricted to Australia. Ann. Ent. Soc. America **59**, 144—147: II (1965) Multiple sex chromosomes. Chromosoma (Berlin) **17**, 323—327.

ONO, T., 1937: On sex-chromosomes in wild hops. Bot. Mag. (Tokyo) **51**, 110—115.

OMACHI, F., and N. UESHIMA, 1957: A study on local variation of chromosome complement in *Scapsipedus aspersus* Chopard (Orthoptera: Gryllidea) Bull. Fac. Agr. Mie Univ. **14**, 43—49.

Omodeo, P., 1951: Problemi genetici connessi con la poliploidia di alcuni lombrichi. Boll. Zool. **18**, 123—129.
— 1952: Cariologia dei Lumbricidae. Caryologia **4**, 173—275.
Östergren, G., and S. Frost, 1962: Elimination of accessory chromosomes from the roots in *Haplopappus gracilis*. Hereditas (Lund) **48**, 363—365.
Owen, J. J. T., 1965: Karyotype studies on *Gallus domesticus*. Chromosoma (Berlin) **16**, 601—608.

Panitz, R., 1965: Heterozygote Funktionsstrukturen in den Riesenchromosomen von *Acricotopus lucidus*. Puffs als Orte uniloker Strukturmutationen. Chromosoma (Berlin) **17**, 199—218.
Partenan, C. R., 1956: Comparative microphotometric determinations of desoxyribonucleic acid on normal and tumerous growth of fern prothalli. Cancer Res. **16**, 300—305.
— I. M. Sussex, and T. A. Steeves, 1955: Nuclear behaviour in relation to abnormal growth in fern prothalli. Amer. J. Bot. **42**, 245—256.
Pätau, K., D. W. Smith, E. Therman, S. L. Inhorn, and H. P. Wagner, 1960: Multiple congenital anomaly caused by an extra chromosome. Lancet i, 790.
Patterson, J. T., and W. S. Stone, 1952: Evolution in the genus *Drosophila*. Macmillan, New York.
Peacock, J., 1965: Chromosome replication. Pp. 101—129 in "Genes and Chromosomes, Structure and Function." Natl. Cancer Inst. Mono. 18.
Pennock, L. A., 1965: Triploidy in parthenogenetic species of the teiid lizard genus *Chemidophorus*. Science **149**, 539—540.
Penrose, L. S., 1963: The biology of mental defect. Sidgwick and Jackson, London.
Perrot, J. L., 1933: La spermatogenèse et l'ovogenèse de *Lepisma (Thermobia) domestica*. Hétéropycnose dans un sexe homogamétique. Z. Zellforsch. **18**, 573—592.
Pierce, W. P., 1937: The effect of phosphorus on chromosome and nuclear volume in a violet species. Bull. Torrey. Bot. Club **64**, 345—356.
Pienaar, R. de V., 1963: Sitogenese studies in "Die Genus *Ornithogalum* L." J. S. Afric. Bot. **20**, 111—130.
Piza, S. de T., 1946: Uma nova modalidade de sexo-determinacao no grilo sulamericano *Eneoptera surinamensis*. Ann. Esc. Agric. Queiroz **3**, 69—88.

Rabl, C., 1885: Über Zellteilung. Morph. Jb. **10**, 214—330.
Randolph, L. F., 1941: Genetic characteristics of the B-chromosomes in Maize. Genetics **26**, 608—631.
Rees, H., 1955: Genotypic control of chromosome behaviour in rye I. Inbred lines. Heredity (London) **9**, 93—116.
— 1964: The question of polyploidy in the Salmonidae. Chromosoma (Berlin) **15**, 275—279.
— and J. B. Thompson, 1955: Localisation of chromosome breakage at meiosis. Heredity (London) **9**, 399—407.
— and S. Sun, 1965: Chiasma frequency and the disjunction of interchange associations in rye. Chromosoma (Berlin) **16**, 500—510.
— F. M. Cameron, M. H. Hazarika, and G. H. Jones, 1966: Nuclear variation in diploid angiosperms. Nature **211**, 828—830.
— and G. H. Jones, 1967: Chromosome evolution in Lolium. Heredity (London) **22**, 1—18.
Resende, F., and P. Da Franca, 1946: Sur l'origine de nouvelles formes II. Portug. Acta Biol. (A) **1**, 289—307.
Revell, S. H., 1947: Controlled X-segregation at meiosis in *Tegenaria*. Heredity (London) **1**, 337—347.
Rickards, G. K., 1964: Some theoretical aspects of selective segregation in interchange complexes. Chromosoma (Berlin) **15**, 140—155.
Riley, R., and V. Chapman, 1961: Origin of genetic control of diploid-like behaviour of polyploid wheat. J. Heredity **52**, 22—25.
Ritossa, F., 1962: New puffing pattern induced by temperature shock and DNP in *Drosophila*. Experientia **18**, 571—572.
— J. F. Pulitzer, H. Swift, and R. C. von Borstel, 1965: On the action of ribonuclease in salivary gland cells of *Drosophila*. Chromosoma (Berlin) **16**, 144—151.
Rothfels, K. H., 1950: Chromosome complement, polyploidy and supernumeraries in *Neopodismopsis abdominalis* (Acrididae). J. Morph. **87**, 287—316.
— and T. W. Fairlie, 1956: The non-random distribution of inversion breaks in the midge *Tendipes decorus*. Canad. J. Zool. **35**, 221—263.

ROTHFELS, K. H., M. ASPDEN, and M. MOLLISON, 1963: The W-chromosome of the budgerigar (*Melopsittacus undulatus*). Chromosoma (Berlin) **14**, 459—467.

RUSCH, M. E., 1960: Untersuchungen über Geschlechtsbestimmungsmechanismen bei Copepoden. Chromosoma (Berlin) **11**, 419—432.

RUSSELL, L. B., 1961: Genetics of mammalian sex chromosomes. Science **133**, 1795—1803.

RUTISHAUSER, A., 1956: Chromosome distribution and spontaneous chromosome breakage in *Trillium grandiflorum*. Heredity **10**, 367—407.

— 1960 a: Fragmentchromosomen bei *Crepis capillaris*. Beiheft zu den Zeitschr. des schweiz. Forstv. **30**, 93—107.

— 1960 b: Zur Genetik überzähliger Chromosomen. Archiv der Julius-Klaus-Stiftung der Vererbungsforsch. **35**, 440—458.

— and E. ROTHLISBERGER, 1966: Boosting mechanism of B-chromosomes in *Crepis capillaris*. Pp. 28—30 in "Chromosomes Today." Vol. 1. Oliver and Boyd, Edinburgh.

SAEZ, F. A., 1932: Variación numérica en correlación con la existencia de cromosomas en *Aleuas viticollis* St. (Orthoptera: Acrididae). Rev. Mus. La Plata **33**, 189—193.

— 1956: Cytogenetics of south american Orthoptera. Nature (London) **177**, 490.

— and C. L. SOLARI, 1959: Chromosome studies in three species of *Scottusa* (Orthoptera: Cyracanthacridinae). Caryologia **11**, 358—367.

SAMEJIMA, J., 1958: Meiotic behaviour of accessory chromosomes and their distribution in natural populations of *Lilium medeoloides* A. Gray. Cytologia (Tokyo) **23**, 159—171.

SANDBERG, A. A., G. F. KOEPF, T. ISHIHARA, and T. S. HAUSCHKA, 1961: An XYY human male. Lancet **ii**, 488—489.

SANDERSON, A., 1960: The cytology of a diploid bisexual spider beetle, "*Ptinus claripes*" Panzer and its triploid gynogenetic form "*mobilis*" Moore. Proc. Roy. Soc. Edinburgh **B 67**, 333—350.

SANNOMIYA, M., 1962: Intra-individual variation in number of A- and B-chromosomes in *Patanga japonica*. C. I. S. **3**, 30—32.

SATINA, S., 1959: Chimeras. Pp. 132—151 in "Blakeslee: The genus *Datura*." Ronald Press, New York.

SCHMIDT, A., 1961: Zytotaxonomische Untersuchungen an europäischen *Viola*-Arten der Sektion Nominum. Öst. bot. Z. **108**, 20—88.

SCHRADER, F., 1941: The spermatogenesis of the earwig *Anisolabis maritima* Bon. with reference to the mechanism of chromosome movement. J. Morph. **68**, 123—141.

— and S. HUGHES-SCHRADER, 1956: Polyploidy and fragmentation in the chromosomal evolution of various species of *Thyanta* (Hemiptera). Chromosoma (Berlin) **7**, 469—496.

— — 1958: Chromatid autonomy in *Banasa* (Hemiptera: Pentatomidae). Chromosoma (Berlin) **9**, 193—215.

SCHUSTER, J., and A. G. MOTULSKY, 1962: Exceptional sex-chromatin pattern in male pseudohermaphroditism with XX/XY/XO mosaicism. Lancet **i**, 1074—1075.

SEILER, J., 1914: Das Verhalten der Geschlechtschromosomen bei Lepidopteren. Nebst einem Beitrag zur Kenntnis der Eireifung, Samenreifung und Befruchtung. Arch. Zellforsch. **13**, 159—269.

— and K. SCHAFER, 1941: Der Chromosomencyclus einer diploid parthenogenetischen *Solenobia triquetrella*. Rev. suisse Zool. **49**, 537—540.

SHARMAN, G. B., 1961: The mitotic chromosomes of marsupials and their bearing on taxonomy and phylogeny. Aust. J. Zool. **9**, 38—60.

— A. J. MCINTOSH, and H. N. BARBER, 1950: Multiple sex chromosomes in the marsupials. Nature **166**, 996.

— and H. N. BARBER, 1952: Multiple sex chromosomes in the marsupial *Potorous*. Heredity (London) **6**, 345—355.

SHERMAN, M., 1946: Karyotype evolution: a cytogenetic study of seven species and six interspecific hybrids of *Crepis*. Univ. Calif. Publ. Bot. **18**, 369—408.

SHINJI, O., 1931: The evolutional significance of the chromosomes of the Aphidae. J. Morph. **51**, 373—433.

SIRE, M. W., and R. A. NILAN, 1959: The relationship of oxygen post-treatment and heterochromatin to X-ray induced chromosome aberration frequencies in *Crepis capillaris*. Genetics **44**, 124—136.

SMITH, B. W., 1955: Sex chromosomes and natural polyploidy in dioecious *Rumex*. J. Heredity **46**, 226—232.

Smith, D., 1965: Transplantation of the nuclei of primordial germ cells into enucleated eggs of *Rana pipiens*. Proc. Natl. Acad. sci. U.S. **54**, 101—107.

Smith, S. G., 1952: The evolution of heterochromatin in the genus *Tribolium* (Tenebrionidae: Coleoptera). Chromosoma (Berlin) **4**, 585—610.

— 1953: Chromosome numbers of Coleoptera. Heredity (London) **7**, 31—48.

— 1956: The status of supernumerary chromosomes in *Diabrotica*. J. Heredity **47**, 157—164.

— 1959: The cytogenetic basis of speciation in Coleoptera. Proc. Xth Int. Congr. Genet. **1**, 444—450.

— 1960: Chromosome numbers of Coleoptera II. Canad. J. Gen. and Cytol. **2**, 66—88.

— 1962 a: Chromosome polymorphism and inter-relationships among bark weevils of the genus *Pissodes* Germar: an ammendment. The Nucleus **5**, 65—66.

— 1962 b: Cytogenetic pathways in beetle speciation. Canad. Entomol. **94**, 941—955.

— 1965 a: Heterochromatin, Colchicine and karyotype. Chromosoma (Berlin) **16**, 162—165.

— 1965 b: Cytological species-separation in Asiatic *Exochomus* (Coleoptera: Coccinellidae). Canad. J. Genet. and Cytol. **7**, 363—373.

— and R. S. Edgar, 1954: The sex-determining mechanism in some N. American Cicindelidae (Coleoptera). Rev. suisse de Zool. **61**, 657—667.

Snoad, B., 1955: Somatic instability of chromosome number in *Hymenocallis calathinum*. Heredity (London) **9**, 129—134.

Soller, M., M. Wysoki, and B. Padeh, 1966: A chromosome abnormality in phenotypically normal Saanen goats. Cytogenetics **5**, 88—93.

Soriano, J. D., 1957: The genus *Collinsia* IV. The cytogenetics of colchicine-induced reciprocal translocations in C. heterophylla. Bot. Gaz. **118**, 139—145.

Sperlich, D., 1966: Equilibria for inversions induced by X-rays in isogenic strains of *Drosophila pseudoobscura*. Genetics **53**, 835—842.

Staiger, H., 1954: Der Chromosomendimorphismus beim Prosobranchier *Purpura lapillus* in Beziehung zur Ökologie der Art. Chromosoma (Berlin) **6**, 419—478.

— 1955: Reziproke Translokationen in natürlichen Populationen von *Purpura lapillus* (Prosobranchia). Chromosoma (Berlin) **7**, 181—197.

— and C. H. Bozquet, 1954: Cytological demonstration of female heterogamety in Isopods. Experientia **10**, 64—66.

Stebbins, G. L., 1958: Longevity, habitat, and release of genetic variability in the higher plants. Cold Spring Harbor Symposia on Quantitative Biology **23**, 365—378.

— 1966: Chromosomal variation and evolution. Science **152**, 1463—1469.

Steffen, K., 1956: Endomitosen im Endosperm von *Pedicularis palustris*. Planta (Berlin) **47**, 625—652.

Stich, H. F., G. L. van Hoosier, and J. J. Trentin, 1964: Viruses and mammalian chromosomes. Chromosome aberrations by human adenovirus type 12. Exp. Cell Res. **34**, 400—403.

Strandhede, S. O., 1965: Chromosome studies in *Eleocharis* Subser. *palustres* III. Observations on western European taxa. Opera Bot. **9**, 1—86.

Stalker, H. D., 1956: A case of polyploidy in Diptera. Proc. Natl. Acad. Sci. **42**, 194—199.

Sugiyama, M., 1933: Behaviour of the sex chromosomes in the spermatogenesis of the Japanese earwig *Anisolabis marginalis*. J. Fac. Sci. Tokyo Univ. iv. **3**, 177—1882.

Suomalainen, E., 1958: On polyploidy in animals. Proc. Finn. Acad. Sci. and Letters **1958**, 1—15.

— 1965: On the chromosomes of the geometrid moth genus *Cidaria*. Chromosoma (Berlin) **16**, 166—184.

— and O. Halkka, 1963: The mode of meiosis in the Psyllina. Chromosoma (Berlin) **14**, 498—510.

Suzuki, S., 1951: Cytological studies on spider I. A comparative study of the chromosomes in the family Agriopidae. J. Sci. Hiroshima Univ. **Bl** (Zool.) **12**, 67—98.

Svardson, G., 1945: Chromosome studies on Salmonidae. Rept. Swedish State Inst. Fresh-Water Fishery Research **23**, 1—151.

Swanson, C. P., 1940: The distribution of inversions in *Tradescantia*. Genetics **25**, 438—465.

— 1957: Cytology and cytogenetics. Prentice Hall, New Jersey.

Takenouchi, Y., 1961: The cytology of bisexual and parthenogenetic races of *Scepticus griseus* Roelofs (Curculionidae: Coleoptera). Canad. J. Genetics and Cytol. **3**, 237—241.

THERMAN, E., 1953: Chromosomal evolution in the genus *Polygonatum*. Hereditas (Lund) **39**, 277—288.

TJIO, J. H., and A. LEVAN, 1954: Chromosome analysis of three hyperdiploid ascites tumours of the mouse. Lunds Univ. Arsskrift NF **2**, **50**, No. 15.

TOBGY, H. A., 1943: A cytological study of *Crepis fuliginosa, C. neglecta* and their F_1 hybrid, and its bearing on the mechanism of phylogenetic reduction in chromosome number. J. Genetics **45**, 67—111.

TROSKO, J. E., and S. WOLF, 1965: Strandedness of *Vicia faba* chromosomes as revealed by enzyme digestion studies. J. Cell Biol. **26**, 125—135.

TSCHERMAK-WOESS, E., 1956: Notizen über die Riesenkerne und "Riesenchromosomen" in den Antipoden von *Aconitum*. Chromosoma (Berlin) **8**, 114—134.

— 1957: Über Kernstrukturen in den endopolyploiden Antipoden von *Clivia miniata*. Chromosoma (Berlin) **8**, 637—649.

UESHIMA, N., 1966: Cytotaxonomy of the Triatominae (Reduviidae: Hemiptera). Chromosoma (Berlin) **18**, 97—122.

ULLERICH, F.-H., 1966: Karyotyp und DNS-Gehalt von *Bufo bufo, B. viridis, B. bufo* × *B. viridis* und *B. calamita* (Amphibia, Anura). Chromosoma (Berlin) **18**, 316—342.

— 1967: Weitere Untersuchungen über Chromosomenverhältnisse und DNS-Gehalt bei Anuren (Amphibia). Chromosoma (Berlin) **21**, 345—368.

— H. BAUER, and R. DIETZ, 1964: Geschlechtsbestimmung bei Tipuliden (Nematocera, Diptera). Chromosoma (Berlin) **15**, 591—605.

UPCOTT, M., 1939: The genetic structure of *Tulipa III*. Meiosis in polyploids. J. Genet. **37**, 303—339.

URSPRUNG, H., and C. L. MARKERT, 1963: Chromosome complements of *Rana pipiens* embryos developing from eggs injected with protein from adult liver cells. Developmental Biol. **8**, 309—321.

UZZELL, T. M., 1963: Natural triploidy in salamanders related to *Ambystoma jeffersonianum*. Science **139**, 113—115.

VANDEL, A., 1940: La parthénogénèse géographique **IV**. Polyploidie et distribution géographique. Bull. Biol. **74**, 94—100.

VED BRAT, S., 1965: Genetic systems in *Allium* I. Chromosome variation. Chromosoma (Berlin) **16**, 486—499.

VIRKKI, N., 1964: On the cytology of some neotropical chrysomelids (Coleoptera). Ann. Acad. Sci. Fenn. IV **75**, 1—30.

— and C. M. PURCELL, 1965: Four pairs of chromosomes: the lowest number in Coleoptera. J. Heredity **41**, 71—74.

WAHRMAN, J., 1954: Evolutionary changes in the chromosome complement of the Amelinae (Orthoptera, Mantoidea). Experientia **10**, 176—177.

— 1966: A carabid beetle with only eight chromosomes. Heredity (London) **21**, 154—159.

— and A. ZAHAVI, 1958: Cytogenetic analysis of mammalian sibling species by means of hybridisation. Proc. Xth Int. Congress of Genetics **II**, 304—305.

WALTON, A. C., 1916: *Ascaris canis* (Werner) and *Ascaris felis* (Goeze): a taxonomic and cytological comparison. Biol. Bull. **31**, 364—371.

— 1924: Studies on nematode gametogenesis. Z. Zellenlehre **1**, 167—239.

WARKANY, J., E. H. Y. CHU, and E. KAUDER, 1962: Male pseudohermaphroditism and chromosomal mosaicism. Amer. J. Diseases of Children **104**, 172—179.

WEILER, C., and S. OHNO, 1962: Cytological confirmation of female heterogamety in the African water frog (*Xenopus laevis*). Cytogenetics **1**, 217—223.

WESTERGAARD, M., 1940: Studies on cytology and sex determination in polyploid forms of *Melandrium album*. Dansk. Bot. Arkiv. **10**, 1—131.

WHITE, M. J. D.: Animal cytology and evolution, 1st edit. 1945, 2nd edit. 1954. Univ. Press, Cambridge.

— The cytology of the Cecidomyidae (Diptera) II (1946). The chromosome cycle and anomalous spermatogenesis of *Miastor*. J. Morph. **79**, 323—369. — III (1947). The spermatogenesis of *Taxomyia taxi*. J. Morph. **80**, 1—24.

— 1950: Cytological studies on gall midges (Cecidomyidae). Univ. Texas Publ. **5007**, 1—80.

— 1951 a: Cytogenetics of Orthopteroid insects. Adv. Genet. **4**, 267—330.

— 1951 b: Structural heterozygosity in natural populations of the grasshopper *Trimerotropis sparsa*. Evolution **5**, 376—394.

White, M. J. D., 1956: Adaptive chromosomal polymorphism in an australian grasshopper. Evolution **10**, 298—313.

— 1957 a: Some general problems of chromosomal evolution and speciation in animals. Survey Biol. Progress **3**, 109—147.

— 1957 b: Cytogenetics of the grasshopper *Moraba scurra* I. Meiosis of interracial and interpopulation hybrids. Australian J. Zool. **5**, 285—304.

— 1961: Cytogenetics of the grasshopper *Moraba scurra* VI. A spontaneous pericentric inversion. Australian J. Zool. **9**, 784—790.

— 1964: Principles of karyotype evolution in animals. Proc. XIth Int. Congress Genetics, 391—397. Pergamon Press, London.

— and N. H. Nickerson, 1951: Structural heterozygosity in a very rare species of grasshopper. Amer. Nat. **85**, 239—246.

— J. Cheney, and K. H. L. Key, 1963: A parthenogenetic species of grasshopper with complex structural heterozygosity (Orthoptera: Acridoidea). Australian J. Zool. **11**, 1—19.

— H. L. Carson, and J. Cheney, 1964: Chromosomal races in the australian grasshopper *Moraba viatica* in a zone of geographic overlap. Evolution **18**, 417—429.

Wickbom, T., 1945: Cytological studies of Dipnoi, Urodela, Anura and Emys. Hereditas (Lund) **31**, 241—346.

Wilson, E. B., 1905: Studies on chromosomes II. The paired microchromosomes, idiochromosomes and heterotropic chromosomes in Hemiptera. J. Exp. Zool. **2**, 507—545.

— 1912: Studies on chromosomes VIII. Observations on the maturation phenomenon in certain Hemiptera and other forms, with considerations on synapsis and reduction. J. Exp. Zool. **13**, 345—431.

Wilson, G. B., and E. R. Boothroyd, 1941: Studies in differential reactivity I. The rate and degree of differentiation in the somatic chromosomes of *Trillium erectum*. Can. J. Res. **19**, 400—412.

Winge, Ø., 1930: Zytologische Untersuchungen über die Natur maligner Tumoren II. Teerkarzinomie bei Mäusen. Z. Zellforsch. u. Mikr. Anat. **10**, 683—735.

— 1940: Taxonomic and evolutionary studies in *Erophila* based on cytogenetic investigations. C. R. Lab. Carlsberg. **23**, 41—74.

Winkler, H., 1916: Über die experimentelle Erzeugung von Pflanzen mit abweichenden Chromosomenzahlen. Z. Bot. **8**, 417—531.

Wolf, B. E., 1956: Nachweis eines lokalisierten Chromatidenstückaustausches im X-Chromosom von *Phryne cincta* mit Hilfe der Speicheldrüsenanalyse. Verhandl. der Deutschen Zool. Gesellschaft in Hamburg 1956, 284—297.

— 1960: Zur Karyologie der Eireifung und Furchung bei *Cloeon dipterum* L. (Bengtsson) (Ephemerida, Baetididae). Biol. Zentral. **79**, 153—198.

— 1961: Y-Chromosom und überzählige Chromosomen in dem polytänen Somakern von *Phryne cincta* Fabr. (Diptera, Nematocera). Verh. deutsch. zool. Ges. Saarbrücken 110—123.

Yamamoto, Y., 1938: Karyologische Untersuchungen bei der Gattung *Rumex*. Mem. Coll. Agr. Kyoto Univ. **43**, 1—59.

Yerganian, G., R. Kato, M. J. Leonard, H. J. Gagnon, and L. A. Grodzins, 1960: Pp. 49—96 in "Cell Physiology of Neoplasia." Univ. Texas Press, Austin.

— and S. Papoyan, 1965: Homomorphic sex chromosomes, autosomal heteromorphism and telomeric associations in the grey hamster of Armenia, *Cricetulus migratorius* Pall. Hereditas (Lund) **52**, 307—319.

Yosida, T. H., and K. Amano, 1965: Autosomal polymorphism in laboratory bred and wild Norway rats, *Rattus norvegicus*, found in Misima. Chromosoma (Berlin) **16**, 658—667.

— A. Nakamura, and T. Fukaya, 1965: Chromosomal polymorphism in *Rattus rattus* (L) collected in Kusudomari and Misima. Chromosoma (Berlin) **16**, 70—78.

Yunis, J. J., 1965: Human chromosomes in disease. Pp. 187—242 in "Human Chromosome Methodology." Academic Press, London.

Zen, S., 1961: Chiasma studies in structural hybrids VI. Heteromorphic bivalent and reciprocal translocation in *Allium fistulosum*. Cytologia (Tokyo) **26**, 67—73.

Zohary, D., and U. Nur, 1959: Natural triploids in the orchard grass, *Dactylis glomerata* L., polyploid complex and their significance for gene flow from diploid to tetraploid levels. Evolution **13**, 311—317.

Species Index

Author Index

Fortsetzung von der 4. Umschlagseite

Band VI/F/1:

The Meiotic System. By **B. John,** Birmingham, and **K. R. Lewis,** Oxford. With 195 figures. IV, 335 pages. 8vo. 1965. S 860.—, DM 136.50, $ 34.15

„...Die vorliegende, mit großer Sachkenntnis zusammengestellte Monographie gibt einen hervorragenden Überblick über den gegenwärtigen Stand des Wissens zum Meiosesystem und seinen Varianten... Das Buch läßt kaum irgendwelche Wünsche offen und ist für denjenigen, der über Fragen der Meiose arbeitet oder sich eingehender über die Meiosesysteme orientieren will, unentbehrlich.“ *Biologisches Zentralblatt*

Band VI/F/2:

Les altérations de la méiose chez les animaux parthénogénétiques. Par **Marguerite Narbel-Hofstetter,** Lausanne. Avec 112 figures (686 dessins ou photographies). IV, 163 pages. in-8°. 1964. S 397.—, DM 63.—, $ 15.75

„...Bei diesem Teil des Handbuches handelt es sich um den gut gelungenen Versuch, durch Zusammenfassung einer Fülle von Einzeltatsachen nach der weit zerstreuten zoologischen Literatur ein abgerundetes Bild vom Stande der Forschung zu vermitteln.“ *Kolloid-Zeitschrift*

Band VI/F/3:

Fortpflanzungsmodus und Meiose apomiktischer Blütenpflanzen. Von **A. Rutishauser,** Zürich. Mit 86 Textabbildungen. V, 245 Seiten. Gr.-8°. 1967. S 540.—, DM 86.—, $ 21.50

Das Buch gibt einen Überblick über das Gesamtproblem der Apomixis. Das Hauptgewicht liegt dabei auf der Darstellung der embryologischen Mechanismen und ihrer Ableitung von jenen der sexuellen Pflanzen, ferner der Meioseanomalien, ihrer Ursachen, ihrer Beziehungen zum Fortpflanzungsmodus und ihrer Bedeutung für die Evolution der Apomikten. Das Buch bringt eine Zusammenstellung der Untersuchungsergebnisse über die genetischen Grundlagen der Apomixis und ihrer Modifikation durch Umweltfaktoren. Ein umfangreiches Kapitel ist der Evolution der Apomikten, der Ausweitung und der Umkombination des Genbestandes apomiktischer Artkomplexe gewidmet.

Band VI/G/1:

The Behavior of Centrioles and the Structure and Formation of the Achromatic Figure. By **H. A. Went,** Pullman. With 30 figures. IV, 109 pages. 8vo. 1966. S 280.—, DM 44.—, $ 11.—

„...Die Fortschritte auf diesem Gebiet haben nur selten ihren Niederschlag in den Lehrbüchern gefunden. Schon aus diesem Grunde schließt die hier vorliegende Monographie eine empfindliche Lücke in der Literatur... Trotz der großen Vielfalt der mitgeteilten Einzelbefunde bleibt die Darstellung stets übersichtlich, klar durchschaubar und gut verständlich.“ *Der Züchter*

In Vorbereitung:

Band VI/B:

The Chromosome Cycle in Mitosis. By **B. John,** Birmingham, and **K. R. Lewis,** Oxford.